Palgrave Insights into Apocalypse Economics

Series Editor
Richard Westra
Centre for Macau Studies
University of Macau
Macau, China

This series is set to become the lodestone for critical Marxist and related Left scholarship on the raft of apocalyptic tendencies enveloping the global economy and society. Its working premise is that neoliberal policies from the 1980s not only failed to rejuvenate capitalist prosperity lost with the demise of the post-Second World War 'golden age' economy but in fact have generated a widening spectrum of pathologies that threaten humanity itself. At the most fundamental level the series cultivates state of the art critical political economic analysis of the crises, recessionary, deflationary and austerity conditions that have beset the world economy since the global meltdown of 2008–2009. However, though centered on work that critically explores global propensities for devastating financial convulsions, ever-widening inequalities and economic marginalisation due to information technologies, robotised production and low wage outsourcing, it seeks to draw on exacerbating factors such as climate change and global environmental despoliation, corrupted food systems and land-grabbing, rampant militarism, cyber crime and terrorism, all together which defy mainstream economics and conventional political policy solutions.

For critical Marxist and related Left scholars the series offers a non-sectarian outlet for academic work that is hard-hitting, inter/trans-disciplinary and multiperspectival. Its readership draws in academics, researchers, students, progressive governmental and non-governmental actors and the academically-informed public.

More information about this series at
http://www.palgrave.com/gp/series/15867

Laura Westra

Climate Change and Starvation

From Apocalypse to Integrity

palgrave
macmillan

Laura Westra
Maple, ON, Canada

ISSN 2523-8108 ISSN 2523-8116 (electronic)
Palgrave Insights into Apocalypse Economics
ISBN 978-3-030-42123-6 ISBN 978-3-030-42124-3 (eBook)
https://doi.org/10.1007/978-3-030-42124-3

Cover illustration: Dina Belenko / Alamy Stock Photo

This Palgrave Pivot imprint is published by the registered company Springer Nature Switzerland AG.
The registered company address is: Gewerbestrasse 11, 6330 Cham, Switzerland

For the brave Indigenous peoples of the forest: the only defenders of the Amazon against political and corporate greed
Laura Westra, 2020

Contents

CHAPTER 1

Introduction

Abstract Westra uses the first chapter to explain the salient points of the situation. Climate change is now viewed widely as a crime against humanity, and the emphasis worldwide is on the power of young people, culminating in the crusade of Greta Thunberg, their spokesperson, but clearly visible through the media in the large peaceful but unyielding protests against the inaction of governments. The latter may agree in words but fail to act, as they pander to voters, be they coal miners, racists and migrant haters and non-white refugees. The central point of the work is the incapable connection between climate change and damages imposed by meat-based diets that inflict as much grave harms on the "haves" as they do on the "have nots", albeit in a different manner. The combination of climate excesses and the ravages of hunger in all its aspects is nothing less than a veritable war, waged against the weakest and forest inhabitants of the world. The results of this war can be seen from the hordes of desperate migrants from South America to North Africa, to the Middle East. The food battle includes the fight against the corporate reliance on imposed GMOs and the toxic aspects of agribusiness. Forests and fields will be the most relevant part of the counteroffensive as will the protection of water and air: what is needed is a global governance that is aware of the reality on the group, not the correct reliance on weakened states unable to do more than close borders and walls against the desperate.

L. Westra, *Climate Change and Starvation*, Palgrave Insights into Apocalypse Economics,
https://doi.org/10.1007/978-3-030-42124-3_1

Keywords Closed borders • Ineffectual states • Greta Thunberg • Young people power • War against the poor

1.1 Climate Change and Starvation: From Apocalypse to Integrity

Climate change is now openly called a crime against humanity, and at least in Europe, it is named as the most basic and important origin of new policies and goals for that continent. Children mount protests almost everywhere, and the originator of the children's movements Greta Thunberg even sailed across the Atlantic to bring her message and that of the children in her movements to the White House. The 2019 G7 meetings debated it as well, while attempting to halt the most obvious sign of its presence, that is, the burning of the Amazon forests, and scientists join in signing declarations that proclaim that reality in all its manifestation while clearly indicting the guaranteed further disasters that await the world as a whole.

Yet each separate government, while proclaiming the truth of climate change in front of news cameras, still hesitates to eliminate their mines, whether at home or even abroad through their corporate citizens, and still continue to offer clear support to conservative political candidates, while denying their obligation to the migrants and refugees who, in order to escape the ravages of climate change, are forced to seek asylum, as melting glaciers, raising water levels, tsunamis, hurricanes, floods, wildfires and weather extremes everywhere confirm the presence of the situation state leaders still deny through their actions.

Equally missing from the main concerns of authorities, even those who are prepared to acknowledge climate change at least in principle, is the question of current diets, specifically the meat-intensive diets that the citizens of the affluent Western countries are led to believe are the only acceptable form of nourishment, while those in the poor countries elsewhere are aiming to achieve the same goal. But the question of food choices, as discussed in this Chap. 1, is too basic to be avoided: it will be the central point of this work.

Food production and consumption, as we will argue, is both a major causative agent of climate change and equally causative of many human rights breaches demonstrated by those affected: the obese. The obese, unhealthy people in wealthy countries, affected by multiple diseases, where cancer now has become the number one reason of death, beyond all cardiac problems. The effects of climate change are equally demonstrated by the starving, desperate refugees and migrants who are described erroneously as "attempting to better their economic conditions", thus compared erroneously with those judged to be more deserving, as victims of a war situation. We will argue that what moves these people makes them essentially, whether it is from South American countries or countries in North Africa or the Middle East' these people are essentially climate refugees, impelled to flee by starvation not the ideal of a future "better economic situation".

In fact the results of Western choices in global governance, especially those related to food choices, are similar to what happens when a country invades another or an internal conflict arises, so the comparison, if not a true identification, with a war is totally appropriate. In addition, the "war" is waged by the wealthier and the stronger, against the weakest and most defenseless in the world. The means of these attacks may not be bombs, guns or other war implements as, we shall argue, the simple reliance on corporate power, especially that which is related to food production, is sufficient.

Of course the generally acknowledged sources of climate change are real and present, and we will examine them in turn, as we will discuss the retreat of the state in the face of the overwhelming power of corporate persons that control our air, water and the food we eat as well as, ultimately, climate itself. These entities are active globally but are not controlled by state laws as they exceed the state in power, nor are they controlled by international law, as they are not "nations", but operate at another level altogether, a level international law has not yet been able to reach.

Hence we can conclude that non-state entities operate without any controls, so the damages they produce, no matter how grave, remain unpunished. This lack of control is particularly evident in the agricultural field, a field that is almost entirely in the hands of major corporations. These non-state entities control or at least direct North American states and extend that control through their monopoly of GMOs; related toxic products, intended to grow fodder; energy substitutes; and all the other products that produce harms from North to South America. Their control is partially extended to Europe, particularly North European countries.

The arguments provided to justify that noxious interference pander to the general ignorance of the public, their trust in their governments' commitment to protect them and their (the agricultural corporations') continued expressed falsehood: theirs, they say, is the only way to provide food for the world's population. We have argued that rather than providing nourishment for the masses, as agribusiness asserts, their products deprive them of nourishment, while exposing them to a great number of physical and mental/emotional diseases from childhood and through to old age.[1]

In fact, the starvation that occurs from South America to North Africa, and which demonstrates their primary and ongoing role in climate change, originates precisely from agribusiness. Science, however, has indicated clearly a number of other sources of climate change, and we will detail each of these below, as they are, one and all, contributing factors; in fact each one has been emphasized in turn as the most significant in recent times. The role played by land and soil and by forests is the main theme discussed in 2019, as it is brought to the public attention by the terrible fires that burn the Amazon region. These forest fires emphasize precisely the main theme of this work: that is the primary role of agribusiness interests as it is the latter that are the main reason for those fires.

Then there is the role of water, equally central to agribusiness and certainly clear in the public eye because most of the obvious effects of climate change involve water, not only as melting glaciers but also as tsunamis, hurricanes and floods. Finally, air that is the basis of climate change is inquinated with constant toxic substances flowing from business operations and visibly changing the temperatures in ways that impose death of the most vulnerable.

In the next chapters, we will start with our personal responsibility defined: do we singly as world citizens and human beings have a real responsibility to work not only against climate change but also against starvation? After all science has been informing us, and knowledge builds first understanding and then responsibility. Science has been speaking clearly about all that is happening in the natural world, not through the authorities and government officials, but through the non-state movements and non-governmental organizations which are not constrained by the political ambitions of state leaders, but are prepared to address clearly the ecological realities on the ground. Lands and forests need to be understood and then respected for their double role, regarding both climate and hunger.

We will also review the issue of water, a basic necessity for agriculture, together with temperature extremes, a clear sign of all the disastrous effects of climate change. These direct climate change effects are combined with the indirect effects that originate from land grabs, which represent a related form of crimes against humanity, an equal contributor to

[1] Westra, L. 2018, On Hunger, International Press, Irvine, CA.

climate change and to starvation as we shall see. It is at this level that the need for global governance is most urgent: our personal responsibility is acknowledged and ongoing, but radical changes in both international and local laws are urgent.

Global governance must acknowledge that the current legal regimes neither understand nor are prepared to approach some of the worst problems faced in almost all continents: the problem of desperate refugees attempting to escape both extreme climactic conditions and the starvation that faces them, as the two conditions are intimately connected. Nevertheless the legal systems, based on faulty understanding of the situation, confuse the means of the escape with the reasons for it. Hence, reacting to the problem with walls on the ground, closed ports on the seashores, rather than with joint national policies of aid and assistance that should transcend the politics of individual countries, is an additional crime in itself.

It is the starvation of these refugees that demands united, responsible and radically new treatment from existing legal regimes, a necessity based upon scientific evidence that is recent and thus exceeds the basis upon which previous legalities were founded. This was recognized by Judge Weeramantry who noted the same occurrences in his 1997 Separate Opinion in the Gabcikovo-Nagymaros Dispute.[2] Judge Weeramantry argued that as the science in 1997 had showed the environmental dangers involved in the project, decisions reached decades earlier could not stand: the law could not prescribe something that represented an attack on human rights, hence something against the principles of law itself.

Thus, the fact that humanity is one, north to south, east to west, now dictates that the scientific discoveries of recent times make the approach to legality of former times not only obsolete but morally wrong: today too, the law cannot and must not prescribe something that represents a direct attack on human rights. For that reason, on the final chapter, we will return to a principle I proposed twenty-seven years ago and have been clarifying and discussing for that length of time with a group of scholars in varied disciplines, united by a common understanding.

References

Westra, Laura, 2018, *On Hunger*, International Press, Irvine, CA

Gabcikovo-Nagymaros Project, IJC Reports 1996, J.C., Weeramantry Separate Opinion 1997

[2] *Gabcikovo-Nagymaros Project*, IJC Reports 1996, J.C., Weeramantry Separate Opinion 1997.

CHAPTER 2

Climate Change and Personal Food Choices

Abstract Westra introduces the multifaceted argument of the work, connecting climate change and hunger and starting at the personal level. Each of us has a responsibility for the ongoing disasters following extreme climate situations which face us. Those who have the time and the interest to research the situation bear the brunt, at least in part, of that responsibility, while many others simply rely on the false/pacifying information released by the administrations of wealthy Western countries. Over the last couple of decades, information has been available to those willing to use the internet to seek it. At the same time, the same sources have fostered our preoccupation with our personal health and fitness. Hence it seems clear that even if our motives for changing our behavior are not driven by a purely moral choice, since the results are such that they bring about the changes that are best not only for our personal health but also, through better diets, best for the world's climate, the move to seek personal health through better diets is a positive one.

Keywords Hunger • Personal responsibility • Diets and behavior changes

L. Westra, *Climate Change and Starvation*, Palgrave Insights into Apocalypse Economics,
https://doi.org/10.1007/978-3-030-42124-3_2

2.1 Introduction

When we consider climate change, the issue is a global one and is so grave that it is easy to believe that only the concerted action by the global community can possibly make a dent into its multiple aspects. Yet, without denying the necessity for such a top-down concerted action, there are growing signs that individual ethical choices may be making a difference.

This chapter intended to discuss the question of personal food choices in relation to climate change, so the first thing to note is that food choices are not only, and perhaps not even, primarily ethical choices, they are choices dictated by our desire to maintain our health at best and, increasingly, our fitness and self-image. In other words, as many facts emerge that were unknown for many decades, except by specialized scientists and institutions, many people now associate food with health, fitness and even the protection of their life.

This change has been slowly but steadily following the information regarding all aspects of food that increasingly targets middle-class people in comfortable situations, perhaps primarily those in larger cities and towns. Hence the phenomena we will discuss sadly do not affect those who desperately need that information, as they are the people whose health and lives are most at risk of the worst effects of climate change: the poor worldwide and, in North America, the communities that still suffer from discrimination, such as African Americans who are still at the mercy of practices of environmental racism.[1]

It would seem that in identifying the populations most affected by climate change, we ought not to be even more strongly committed by engaging in ethical action in all our choices. However, I will argue that the roots of the concern for food choices on the part of those who can afford to choose are less purely moral actions than one might think. In fact those who have both access to the information that should inform their choices and the financial ability to make such choices are primarily women, though men are now part of that group as well, where the desire for self-improvement, combined with optimal fitness in all its health-related aspects, is the main concern, even commitment. The feminist movement has done nothing, ultimately, to displace the ascendant that the media have on the young and not so young women, whose models to emulate remain, at worst, cadaverous-looking models from fashion shows and

[1] Bullard, Robert,1994, *Dumping in Dixie*, Westview Press, Boulder, CO.

various ads, even Barbie dolls: these are, at best, physical ideals that include weight control and health-related fitness.

This aspect of modernity, without exaggeration, is one of the most prevalent ones of our times, again, primarily in larger cities and towns, and at least among the people we have described: those who have both the financial and the educational ability to absorb the ever-mounting flow of information that reaches society. That flow of information is increasing from many sources despite the efforts of the major corporate to powers that have been controlling our food sources all its aspects for too many decades, although they are now facing severe damages in the courts. Those powerful legal persons do not respond to appeals to human rights, let alone to environmental concerns, but they do respond readily to economic hits.

As the undeniable facts emerge regarding their practices, those corporations' actions have been shown to be grossly immoral, though not openly illegal, and finally the courts have started reacting clearly to the effects of those corporate products that have harmed both individuals and environments with decisions in favor of the plaintiffs. The success of those cases have forced civil society as a whole to examine the reasons why the harmful facts have been hidden for so long, as the very basis of the trust they had in their government institutions designed to protect the public has been misplaced in favor of maintaining a relation of complicity with the harmful corporate entities themselves.

We will discuss this sequence of events and their effects in some detail, as well as the intimate relation between these events and climate concerns. Indeed we will show that many of the correct choices that are favorable to radical changes intended to ameliorate climate issues are becoming increasingly desirable and acceptable. Nevertheless, welcome as those choices are, there is as much enlightened self-interest in altered food preferences as there is direct concern with ethical choices.

2.2 The Question of Health and Fitness: The Issue of Our Times

This chapter is intended to discuss the interface between food choices and the concern for climate change. There is no question that recently the people who can afford to follow their food preferences clearly choose food that is good for the environment, and it is a matter of social status more than age. This approach is primarily true in the case of women, although men are also affected: we find women from the mid-twenties to the mid-fifties and beyond, as well as men roughly in the same age group who can be identified in this group. They faithfully populate the gyms, often daily or at least several times a week; discuss extreme diet choices, weights and measurements; the respective abilities of gym instructors; and—of course—various forms of rejuvenation.

This description fits primarily, and perhaps exclusively, persons in cities and larger towns and who are financially comfortable. They travel often, and their faithful companion is the ubiquitous cell phone, so that they can take endless selfies at their chosen destinations: their eyes on their cell phones far more often than on the monuments or landscapes they traveled to enjoy.

What this rather common scenario depicts is the basic mind-set of the group of Western citizens that are most likely to have heard the latest environmental scientific information and can be expected to be concerned about climate change. The question that arises at this point, however, is what really moves this large, interested and well-informed community, what causes them to choose, to act in one way rather than another.

Perhaps we can term these persons part of the "me-now" generation, not to be confused with the "me-too" group that are known to react to sexual misconduct in the protection of their rights, rather than to be motivated by environmental concern. The "me-now" group instead couples the main concern with self-improvement, health and fitness with environmental concerns. This combined preoccupation unfolds against an admiring background of peers via various forms of social media.

Most women feel pressured to lose weight, to get fit, to eat well, to exercise enough (or more), in addition to their family and work obligations, so that the wellness goals may well become impossible, ever-elusive goals. Increasingly, men as well feel the need for "improvement" in order to keep up with their peers, including the younger ones, in the quest for youth of appearance at least.

Of course all their choices regarding food are healthy and thus environmentally friendly, and they can be the best choices to help fight climate change. Healthy food choices—the topic of this chapter—are a decided improvement over diets that are based on older "food guides" in North America, essentially based on business-as-usual, promulgated by governments intent on supporting their industrial agricultural partners, not citizens' health, let alone environmental protection.

In fact, the proliferation of no-meat hamburger, a recent development in most fast-food chains, is made to appeal primarily to the health/fitness requirements of the generation/community of which we have been discussing. They are designed to avoid red meat, laced with hormones and grown through the fodder common to all industrially raised animals, laced with genetically modified feed together with the pesticides increasingly known to be both harmful to human health and contributing to climate change at the same time.[2]

Yet, when you read the description/evaluation of these no-meat or beyond-meat patties, the evaluation reflects the calorie count of each and the taste/consistency appeal. The environmental/climate change component is real, but it remains implicit at best.

The most recent IPCC Report[3] addresses both climate change and meat consumption for the first time explicitly, and we will report the mounting scientific evidence on both environmental issues and public health in the next section. The present question, however, is whether those who are aware of the scientific evidence today actually relate it directly to their food choices.

In simple terms, are the changed food choices truly ethical choices related to climate change? Hence we need to examine what constitutes a moral choice, in order to be able to answer. Kant states:

> Every human being has a legitimate claim to respect from his fellow human beings and is in turn bound to respect every other. Humanity itself is a dignity; for a human being cannot be used merely as a means by any human being…but must always be used at the same time as an end.[4]

Thus dignitarianism is the only position that will provide the unequivocal "no" needed for the defense of the basic right to survival of the most vulnerable people, surely the correct appellation to define those in the third world and those who are marginalized even in wealthy countries, that is, the poorest of humanity.

[2] Westra, L. *On Hunger*, 2018, Brown-Walker Press, Irvine, CA; Gillam, Carey, 2017, Whitewash (The Story of a Weed Killer, Cancer and the Corruption of Science), Island Press, Washington, DC.

[3] IPCC Report 2019; httpsawww.dw.com/en/ipcc-report the-world-gets-hungrier-but-the-land-ls-exhausted-from- us-and-from-climate-change/a 49783271

[4] Kant, Emmanuel, *The Metaphysics of Morals.*

Thus, to continue to eat meat, blithely ignoring how it is produced, how much desperately needed water that production required, is indeed to use large portions of humanity as means to the personal satisfaction of others. Roger Brownsword explains:

> Dignitarianism, it cannot be emphasized too strongly, is a red light, not an amber light ethic, its credo is that we should not proceed at all (where activities are judged categorically to compromise human dignity), rather than we should proceed with care.[5]

If duties are owed to all beings capable of sentience and agency, then the duty is non-specific, but it can be owed to all life and its preconditions—that is, also to the habitats whose fittingness supports our own. In other words, by extending the meaning of dignity from its modern sense of (Kantian) *dignitas* to its classical Greek sense of "within the natural laws of the universe", Kant's imperatives can be also placed within the more far-reaching imperatives of the principles of integrity.[6]

For Kant an objective principle, which is obligatory for a willing agent, is a command of reason, termed an imperative; it is intended as a good, not as something pleasant, but good in the sense that it is so for every human being. Hence, after our discussion of what I termed the "me-now" communities of extremely diet-conscious people, it is not clear that we can answer in the affirmative the question of whether their diet choices are entirely, or even primarily, ethical choices. Perhaps we should admit that those choices are indeed favorable to both the environment and in line with climate-change concerns but that those choices are, at best, only indirectly related to a moral choice.

Perhaps Kantian reasoning, though basic to morality, may remain somewhat remote from the environmental connections we are seeking to tract. What is the interface between dignity and ecological concern?

If we look at Latin sources, most passages using *dignitas* use it just in that sense, although a few exist where a different use can be found, in Cicero and Apuleius, for instance. In *The Golden Ass*, Lucius Apuleius speaks of "French mares" possessing *dignitas*. It is based, however, on

[5] Brownsword, Roger, 2008, *Rights, Regulations and the Technological Revolution*, Oxford University Press, Oxford, p.10); see also Sachs, Wolfgang, 2004, "Climate Change and Human Rights" in *Scripta Veria*, vol.106, *Interactions between Global Change and Human Health*, Pontifical Academy of Sciences, 349–068, Rome, Vatican City.

[6] Westra, Laura, 2016, *Ecological Integrity and Global Governance*, Routledge, London, UK.

their "long antecedents" or their parentage—still a socially based meaning, extended for the occasion.[7]

Yet if we reach even further back to Greek terminology and meaning, we find a much more useful connection. First of all, "dignity" has the dictionary senses of both "value" and "worth". This of course does not help, because it leads to a circular argument: we need to respect persons or ecosystems because they possess "inherent worth". This is based on their innate "dignity" ("worth" or "value")—a circular argument which tells us absolutely nothing of this obscure property which must elicit our respect. To interpose "rationality", "humanity" or "sentience" does not help, unless the argument is consequentialist and utilitarian instead. We simply need to ask why is "rationality" worthy of respect, and what gives it "dignity". Once again, we encounter a dead end.

It is only when we turn to the Green term (*semnotes*) that we find a further meaning: "fittingness". Xenophon, for instance, claims that "it is not fitting for Zeus to move around from place to place" using his limbs, like ordinary mortals, and that is why he moves through his mind. If we can accept conceptually this link between "dignity" and "what is fitting" or "fittingness", then we have a rather solid basis. In requiring reverence or respect for all that is "fitting" in an environmental context, it permits inclusion not only of live individual entities, species and ecosystems but also of the laws that govern their interaction. We no longer need to view all natural intercourse as a bloody "drama" we are "saved from", as Schweitzer does, nor need we feel—as he does—guilt when we give in to "necessity" and force ourselves to kill some living thing, say a carrot or a fish. The painful, paradoxical dichotomy Schweitzer experiences is thus resolved.

The apparent divergence between "fitting" in the social sense, as exemplified by the concept in ancient Greek works and Roman sources, and the environmental "fit" we wish to defend is reconciled through the "natural law" connection. In more recent times, what is "fitting" to a ruler, for instance, is understood in terms of social conventions; in the sense cited from ancient sources, "fittingness" is as natural as the succession of seasons or the freezing of water at low temperatures. Hence the comparison is more accurate than would appear at first glance.

[7] Lucius Apuleius, *The Golden Ass, (Metamorphoses)* Book X, 18, Loeb Classical Library, W. Adlington tr. (1566) revised by S. Gesin (London, UK: Heinemann, 1965).

Today, and in our sense of what "fits", continued ongoing scientific investigation of these natural laws is vital. At the same time, the status and dignity of these laws and processes, as well as that of the natural world itself, which functions according to them, must be upheld. Without expecting to fully connect to the complexities of the Kantian principles or to the background of the dignitarian position in ancient thought, today's enlightened food choices may simply be based on the sense of responsibility, in the sense advocated by Hans Jonas.[8]

Jonas argues that the nature of human action has changed and that the prevailing ideology of the age, the "scientific view of nature", is no longer acceptable for a viable ethic, if it ever was. All previous ethics, he claims, were flawed in regard to our main problem, that of the future. In brief, all previous ethics (a) treated the non-human world as ethically neutral and the human techne connected with the world in the same vein; (b) were anthropocentric; (c) considered man's essential being as unchanging with the result that the mutual impact of the man/techne was not explored; (d) and viewed all "ends" of man as "proximate", both temporally and geographically. Present circumstances demand a changed worldview. Once we realize that when we interact with the non-human world we are—at the same time—affecting not only specific human beings but, a fortiori, a humanity as such, the claim of ethical neutrality can no longer be sustained. The main change which Jonas perceives is in the import of human power. However, great power need not be condemned as such, I think, unless it is coupled with self-serving, controlling arrogance, the hubris of earlier times. Yet as far as we are concerned, up to now there has been, as David Ehrenfeld puts it, an "assumption of omnipotence" which has served as the basis for the "arrogance of Humanism". Ehrenfeld suggests that we need to moderate both our arrogance and our belief that if only we dissect, analyze and fund one problem enough, the "solution" will be found, and all will be well.

And we do not really need to appeal to scholarly works today, to better understand that a position of arrogant self-assurance will not appeal as a solution with any of the major environmental problems that face us, especially climate change. It is sufficient to consider our responsibility to the

[8] Jonas, Hans, 1984, The Imperative of Responsibility, University of Chicago Press, Chicago.

people living on sinking islands, to those who have lost everything because of fires out of control, to the victims of tsunamis, hurricanes and the like.[9]

As long as we are aware of the disasters all around us and of the effects of extreme weather wherever we are, our responsibility is to do whatever is in our power to attempt to halve, if not reverse, the source of all these ills. Yet for the most part, we tend to trust our governments and institutions to take care of such problems; hence to assume that responsibility, despite their obvious failures to do so is primarily theirs. Nevertheless not taking a strong position against what constitutes a clear abuse against the rights of their own citizens renders one complicit in the abuse.

The subject of psychological orientation in various groups and of their views of government and public policy is not intended to exonerate citizens from responsibility, be they corporate or not. A paper by Judith Lee Kissell (1999) analyzes complicity as a multifaceted concept. "Complicity" includes "encouraging", "enticing", "enabling", "ordering" and "failing to intervene", and one can cite examples from antiquity to the present that all fit loosely under the general heading of complicity.

Kissell says:

> For example, we count as accomplices Aeschylus/Aegisthus, who *encourages* Clytemnestra to kill her husband, Agamemnon; Shakespeare's Jago, who *entices* Othello to kill his beloved Desdemona; the mother who *enables* her child to become an alcoholic; the gangleader who *orders* a beating of a victim; the Western powers who, according to Margaret Thatcher, were complicit for *failing to intervene* in the former Yugoslavia.[10]

These examples demonstrate the wide latitude we accord to the concept of complicity in a variety of settings.

We can now turn to our main focus in this section: what constitutes a conspiracy where individual responsibility is present, even in collective actions. The first thing to note is that participation in a "common plan" or "conspiracy" is sufficient to ensure one's responsibility as an individual.

Kissell defines conspiracy as: "an offense in which one agent, the accomplice, becomes responsible for the acts of, and the harm caused by another agent, the perpetrator".[11] Following this definition, she adds: "complicity

[9] Ehrenfeld, David, 978, *The Arrogance of Humanism*, Oxford University Press, Oxford.

[10] Kissell, Judith Lee,1999, "Causation: the Challenge of Complicity", Paper presented at the Eastern American Philosophical Association (one file with author).

[11] Kissell (1999: 2).

is an *offense* and not simply a collaborative action". Nevertheless it is clear that even "planning", "encouraging" or even "enabling" is not in and of themselves harming anyone, when just two are involved, the accomplice and the perpetrator. It seems as though the situation is totally different when a group is involved. The one who delivers a hate speech to a group cannot claim innocence, when the inflamed group acts violently in consequence of hearing the encouragement to hate. The speaker cannot just claim he did not participate in the violence and stood aside from it.[12]

But the questions raised about causation may be more nuanced to help us understand the situation. H.L.A. Hart says:

> The notion that a cause is essentially something that interferes with or intervenes in the course of events which should normally take place is central to the common sense concept of a cause.[13]

If this definition is accepted, then one individual's actions may not influence events in a society, but the aggregate activities or the social action of many may *indeed* make a difference. A case in point may be the European Community's ban of English beef and other meats, coupled with the adamant rejection of biotechnology in agricultural products. My rejection of beef, or yours, while still being a contributing cause to the difficulties Monsanto has encountered in the European market, did not really make a difference in the reduction of the arrogant assumptions of compliance normally taken by Monsanto in their violent activities against all life-forms, natural systems and ultimately all human beings affected by their products.

We can conclude that even though we are not directly responsible for climate change, moral action is required on our part or—minimally—action that reflects in some measure our responsibility to help redress the collective culpable negligence that has led to the present impasse with the complicity of our joint inertia. We, the citizens, are not guilty of allowing climate change to arrive at the present extreme point, but, once scientific research has disclosed the multiple causes that have contributed to its development and once it has even identified the true impact of food choices, the latter can no longer be treated as neutral actions that simply

[12] Hart, H.L.A. 2008, *Punishment and Responsibility* 2nd ed., Oxford University Press, Oxford UK.

[13] Hart and Honoré.

reflect our preferences. Whatever the main reasons that impel us to action, action that reflects our awareness appears to be unavoidable.

In the last section, we mentioned the flow of information from scientific sources available to those who are interested and able to act upon their choices. Both earth and climate scientists and public health specialists have been reaching a public that no longer relies on its own government sources for information and protection from harm but has turned to the internet not only for information but especially for the possibility of joining others of similar interests and concerns through NGOs and other non-state actors.[14]

In this sense, the proliferation of social media is serving an extremely useful function. In North America, especially in the US, a country that strongly resents both regulatory regimes and mandated social support, while giving primacy and control to its corporate/political support groups instead, that trend represents a form of liberation but also an obligation to avoid complicity in the willful blindness of most government sources. The result of this trend is becoming more evident almost each day.

In Manhattan, for instance, diners attract clients through large signs in their windows, "NO GMOs",[15] the proliferation of vegetarian/vegan restaurants and even the arrival of no-meat offerings in the standard large chains, the same companies that used to pride themselves on their American beef, as discussed in Section 2. What is particularly striking and encouraging is that these popular appeals are no doubt intended to increase the profit of those commercial enterprises, as all these novel initiatives fly in the face of the government-approved food guides.

Recently Sheila Jasanoff (Pforzheimer Professor of Science and Technology Studies, Harvard Kennedy School) spoke at a workshop at Erasmus University at Rotterdam.[16] Jasanoff reported that in the US government, sources continue to repeat the same misleading food guide and

[14] Sassen, Saskia, 2006, *Territory, Authority, Rights: From Medieval to Global Assemblages*, Princeton University Press, Princeton, NJ.

[15] Chai-Lin "Fight against GMOs in Taiwan", 2019, unpublished ms. on file with the author.

[16] June 6, 2019, "The Science and Politics of Glyphosate": the workshop included scientists, government officials from both the Netherlands and Europe and the representatives of IARC and Collegium Ramazzini, as well as EFSA, attacked by all for its immoral and illegal practices. Over sixty people were present, including this author as a speaker, and a silent representative of Bayer, who heard the heavy facts against glyphosate, the main ingredient in Monsanto/Bayer's herbicide Roundup.

ignore the mounting court cases and their decisions in favor of those who are harmed by Bayer's products.

The people themselves, especially in the cities, ignore the government sources, as they use NGOs and other sources and turn to buying organic products in unprecedented numbers, rejecting anything that contains GMOs, hence generally use their buying power to follow choices that are environmentally sound and thus helpful to the cause of climate change. Many small and even large companies have abandoned Monsanto/Bayer's products, not for environmental or human rights concerns but in order to protect their profit margin and to avoid the huge risk of losses that now face Bayer: Walmart no longer stocks Roundup herbicide.

How did these changes happen? The climate change connection has made itself felt, not only through appeals to suffering polar bears but through an increase in violent storms, tsunamis, tornadoes and wildfires, plus the increasingly unpredictable temperature changes in all countries.[17]

2.3 The Accelerating Flow of Scientific Information on the Connection Between Food and Climate Change

The increasingly hot temperatures together with the opposite city extremes, bringing unprecedented African heat to North Europe and to most cities in North America, and uncontrollable fires from California to Portugal, but also as far as Siberia, have brought home to all, except those who have political reasons to continue with an obviously false ideology, that humanity is on a dangerous slippery slope, one that requires immediate radical changes, to slow down—if possible—the fall to a climate where only the wealthy, with access to air conditioning, may still survive for a while.

The scientific environmental information has been following two separate streams: the earth, forests and land that need preserving, on one hand, and the protection of species, increasingly decimated each year, on the other.[18]

[17] World Scientists Warning of a Climate Emergency, William J. Ripple, Christopher Wolf, Thomas M. Newsome, Aug.11, 2019 viewed at http://scientistswarning,forestry.oregonstate.edu/

[18] Westra, L., 1994 *The Principle of Integrity*, Rowman Littlefield, Lanham, MD; Pimentel David et al. "Conserving Biological Diversity in Agricultural/Forestry Systems" 42(5)

In 1991, the US Clean Water Act proclaimed the necessity of protecting the biological integrity of water, and in 1987 the *Great Lakes Water Quality Agreement* was signed by both the US and Canada, to ameliorate the poor conditions of the Great Lakes and protect their wildlife.[19] As far as food is concerned, aside from the fact that the WHO and other related European sources had been publishing material linking food and public health for decades, the first clear clarion call for radical changes in food production and distribution came from the 2014 Final Report of Olivier de Schutter.[20]

De Schutter said:

> The normative content of the right to food can be summarized by reference to the requirements of availability, accessibility, adequacy and sustainability, all of which must be built into legal entitlements and secured through accountability mechanisms.[21]

Essentially, Rapporteur de Schutter proposed a radical reform, thus moving beyond capitalism to ensure the fulfillment of the right to food and mentioning the previous Special Rapporteur on the Right to Food in 2001, which stated unequivocally that "Anyone dying from Hunger was dying from murder".[22]

De Schutter links present agricultural practices explicitly to climate change:

> The most potentially devastating impacts of industrial modes of agricultural production stem from their contribution to increased greenhouse gas emissions. Together field-level practices represent approximately of total human made greenhouse gas emissions, in the form of nitrous oxide (N20), from

Bioscience, 1992; Reed, W.E. and M. Wackernagel, 1996, *Our Ecological Footprint*, New Society Publishers, Gabriola Island, BC; Ulanowicz, Robert, 1997, *Ecology, The Ascendent Perspective*, Columbia University Press, NY; Noss, Reed F. and Cooperrider, A. Y., 1994, *Saving Nature's Legacy*, Island Press, Washington D.C.

[19] *Great Lakes Water Quality Agreement of 1978 Between the United States and Canada*, signed at Ottawa Nov.22, 1978, Amended by Protocol signed Nov.18, 1987.

[20] O. de Schutter, 2014, *Final Report: The Transformative Potential of the Right to Food, UN doc.GE.14–10,537*, Human Rights Council 25th Session; FAO 2013, *The State of Food Insecurity in the World 2013: The Multiple Dimensions of Food Security*, FAO, Rome.

[21] de Schutter, "ibid., "Diagnosis", p.),also citing the ICESCR General Comment No. 12.

[22] Ziegler, J., 2002, Report of the Special Rapporteur of the Commission on Human Rights on the Right to Food. Joo. A/57/536, United Nations, New York.

> the use of organic and inorganic nitrogen fertilizers, methane (CH4) from flooded rice fields and livestock, and carbon dioxide (CO_2) from the loss of soil organic carbon in cropland and, due to the intensified grazing on pastures. In addition, the production of fertilizer, herbicides and pesticides, the tillage, irrigation and fertilization and the transport, packaging and conservation of food require considerable amounts of energy, resulting in an additional 15 to 17 per cent of total man-made greenhouse gas emissions attributable to food systems.[23]

Thus the one person charged to produce an in-depth study of the interface between food and climate change confirmed that relation, as "livestock is found to be responsible for 51 per cent of anthropogenic gas emissions".[24] The unavoidable connection between food systems and climate change is confirmed also by the abundant material now available on land grabs, that is, on the appropriation of Indigenous lands by various means, many illegal and all morally wrong, for the purpose of fostering profitable monocultures, mostly intended to grow fodder for animals and material intended to feed automobiles, so neither humans nor animals.[25]

Local Indigenous communities have a right to their natural resources, but the corporate land grabbers, for the most part, also work with the support of the IMF or the World Bank.[26]

The evidence continues to emerge regarding the connection between ecological harms, specifically climate change and food production and consumption. It seems to have accelerated in the last couple of decades, and it discloses the primary culpability of states and corporate interests, together with the clear proof of the manifold harms that result. Hence it seems that everyone ought to accept some degree of moral responsibility for one's food choices as well.

[23] de Schutter, 2014, A/HRC/25/57, No. 7, P-5; see also High Level Panel of Experts on Food Security and Nutrition, *Food Security and Climate Change*, HLPE Report No.), (June 2012)1 Vermeulen S., B. Campbell and J. Ingram, "Climate Change and Food Systems" *Annual Review of' Environment and Resources*, vol. 37 (2012), pp. 195–222.

[24] de Schutter, ibid., point 9, p. 6; Goodland, Robert and Jeff Anhang, "Livestock and Climate Change: What if the Key Actors in Climate Change are Cows, Pigs and Chickens?" World Watch Nov/Dec. 2009.

[25] Liberti, Stefano, *Land Grabbing*, Verso Publishing, London 2013; Pearce, Fred, *The Land Grabbers: The New Fight over Who Owns the Earth*, 2012, Beacon Press, Boston, MA.

[26] MacKay, F, 2005 "The Draft World Bank Operational Policy 4.10 on Indigenous Peoples Progress or More of the Same?" *Arizona Journal of Intl. Law*, Vol. 80, no. 2, pp. 305–330.

2.4 Concluding Thoughts

In the previous section, we discussed some of the major sources of climate change, as well as its worst effects. No doubt the food industries as they function today are the major sources of disasters, as has been witnessed in the fire that is engulfing the largest major forest in the world, the Amazon. It is that region that witnesses the disastrous effects on the local Indigenous populations, not only in Brazil but in the neighboring nations, and the desperate flight of refugees toward the north.

Although North African regions are not equally in the public eye, the situation there is a replica of that taking place in the Amazon: unlivable conditions following land grabs, leading to desperate migrants and refugees; we can therefore conclude that there is a decided responsibility on our part to ensure that our food choices reflect the newly discovered realities. The great majority of people in North America have neither the time nor the interest to seek out the information that support that responsibility.

In contrast, those who are perhaps more free to pursue their interests tend to be at least as preoccupied with their own personal health and fitness as they might be with planetary issues. However, we are not attempting to establish the soundness of peoples' philosophical reasoning, but to establish whether those who view themselves as trendsetters are in fact changing their dietary habits in a way that is constant with the responsibility imposed by the urgency of climate change.

Since the commitment to self-improvement demands food choices that are environmentally sound than uninformed choices have been, the fact that such choices may not be purely moral (in the Kantian sense) and disinterested should not render those choices any less desirable and praiseworthy.

References

Apuleius, Lucius, *The Golden Ass, (Metamorphoses)* Book X, 18, Loeb Classical Library, W. Adlington tr. (1566) revised by S. Gesin (London, UK: Heinemann, 1965)

Brownsword, Roger, 2008, *Rights, Regulations and the Technological Revolution*, Oxford University Press, Oxford, p.10

Bullard, Robert, 1994, *Dumping in Dixie*, Westview Press, Boulder, CO

Chai-Lin "Fight against GMOs in Taiwan", 2019, unpublished ms. on file with the author

de Schutter, O., 2014a, A/HRC/25/57, No. 7, P-5

de Schutter, O., 2014b, *Final Report: The Transformative Potential of the Right to Food, UN doc.GE.14-10537*, Human Rights Council 25th Session

Ehrenfeld, David, 1978, *The Arrogance of Humanism*, Oxford University Press, Oxford

FAO 2013, *The State of Food Insecurity in the World 2013: The Multiple Dimensions of Food Security*, FAO, Rome

Gillam, Carey, 2017, Whitewash (The Story of a Weed Killer, Cancer and the Corruption of Science), Island Press, Washington, DC

Goodland, Robert and Jeff Anhang, "Livestock and Climate Change: What if the Key Actors in Climate Change are Cows, Pigs and Chickens?" World Watch Nov/Dec. 2009

Great Lakes Water Quality Agreement of 1978 Between the United States and Canada, signed at Ottawa Nov.22, 1978, Amended by Protocol signed Nov.18, 1987

Hart, H.L.A. and Honore, T. 1985, Causation in the Law, Clarendon Press, Oxford, UK

Hart, H.L.A. 2008, *Punishment and Responsibility* 2nd ed., Oxford University Press, Oxford UK

High Level Panel of Experts on Food Security and Nutrition, *Food Security and Climate Change*, HLPE Report, (June 2012)

ICESCR General Comment No. 12

IPCC Report 2019; https://www.dw.com/en/ipcc-report-the-world-gets-hungrier-but-the-land-is-exhausted-from-us-and-from-climate-change/a-49783271

Jonas, Hans, 1984, The Imperative of Responsibility, University of Chicago Press, Chicago

Kant, Immanuel, 1964, Groundwork of the Metaphysics of Morals, trans. H. J. Paton, Harper Torchbooks, N.Y.

Kissell, Judith Lee, 1999, "Causation: the Challenge of Complicity", Paper presented at the Eastern American Philosophical Association (one file with author)

Liberti, Stefano, *Land Grabbing*, Verso Publishing, London 2013; Pearce, Fred, *The Land Grabbers: The New Fight over Who Owns the Earth*, 2012, Beacon Press, Boston, MA

MacKay, F, 2005 "The Draft World Bank Operational Policy 4.10 on Indigenous Peoples Progress or More of the Same?" *Arizona Journal of Intl. Law*, Vol.80, no.2, pp.305–330

Noss, Reed F. and Cooperrider, A. Y., 1994, *Saving Nature's Legacy*, Island Press, Washington DC.

Pimentel, David et al. "Conserving Biological Diversity in Agricultural/Forestry Systems" 42(5) *Bioscience*, 1992

Reed, W.E. and M. Wackernagel, 1996, *Our Ecological Footprint*, New Society Publishers, Gabriola Island, BC

Ripple, William J., Christopher Wolf and Thomas M. Newsome, World Scientists Warning of a Climate Emergency, Aug.11, 2019 viewed at http://scientistswarning.forestry.oregonstate.edu/

Sachs, Wolfgang, 2004, "Climate Change and Human Rights" in *Scripta Veria*, vol.106, *Interactions between Global Change and Human Health*, Pontifical Academy of Sciences, 349–068, Rome, Vatican City

Sassen, Saskia, 2006, *Territory, Authority, Rights: From Medieval to Global Assemblages*, Princeton University Press, Princeton, NJ

Ulanowicz, Robert, 1997, *Ecology, The Ascendent Perspective*, Columbia University Press, NY

Vermeulen S., B. Campbell and J. Ingram, "Climate Change and Food Systems" *Annual Review of' Environment and Resources*, vol.37 (2012), pp 195–222

Westra, Laura, 1994, *The Principle of Integrity*, Rowman Littlefield, Lanham, MD

Westra, Laura, 2016, *Ecological Integrity and Global Governance*, Routledge, London, UK

Westra, Laura, 2018, *On Hunger*, Brown-Walker Press, Irvine, CA

Ziegler, J., 2002, Report of the Special Rapporteur of the Commission on Human Rights on the Right to Food, Joo. A/57/536, United Nations, New York

CHAPTER 3

Ecological Integrity and Climate Change

Abstract We start to what turns out to be the most important aspect of both climate change and strain: the scientific findings that increasingly have demonstrated that undeniable connection. Part 1 addresses the role of forests and cultivated lands, while Part 2 deals with water and air issues. The starting point can be traced to the work of Aldo Leopold, a forester, who posited the foundation role played by forests, as he decries the habit of treating nature as property with no regard for the integrity of the Earth. His words have been echoed and further developed through the decades, as one scientist after the other noticed that agribusiness led to soil erosion, soil and water pollution and ecological deterioration. Unexpected support for this position comes also from Canadian/Ontario religious groups such as the Mennonites and Dutch Reformed people, who demand support for the Earth's life-giving and life-supporting capacities, based on their faith's obligation to God's creation, as far back as the early 1990s. A similar position can be found today in the work of Pope Francis, in his impersonate defense of the integrity of the Earth. Finally the clearest explicit connection between climate change and what I have termed Hunger came with the explicit FAO Report of Olivier de Schutter. Respect for nature and its ways is also in ounce in the lifestyle and beliefs of Indigenous peoples, among the most affected by the war on the weakest and the poorest people. Part 2 turns to water, an absolute necessity of life, but exploited in too many ways, despite the sting legal regimes for its protection, be it in lakes, streams or oceans, the former polluted and often unable to support life

L. Westra, *Climate Change and Starvation*, Palgrave Insights into Apocalypse Economics,
https://doi.org/10.1007/978-3-030-42124-3_3

within them, the latter clogged by the plastic that affects all marine life. Finally the air is also unsafe, unhealthy and so polluted that in some countries wearing masks for protection has become a way of life.

Keywords Forests • Productive land • Science in support of climate change/starvation • Polluted water • Unsafe air

3.1 Part 1: The Role of Lands and Forests

We will approach the science of land/agriculture and forestry from the joint standpoint of climate change and starvation, both of which were not some years back the burning issues they are today, when the scientific information has been steadily emerging. Yet, as we shall see, even at the very early stages of research, the issues discussed lay the ground for the present situation. In fact, the basics for what today is commonly viewed as a crime against humanity (e.g., climate change), and the overwhelming presence of harms originating from the reduction of food and from scarcity, actually emerged a long time ago.

Aldo Leopold first published his "Conservation Ethic" in 1933.[1] Leopold addresses the extension of ethics which he terms "a process in ecological evolution":

> The first ethic dealt with the relationship between individuals. The Mosaic Decalogue is an example. Later accretions dealt with the relationship between individual and society. Christianity tries to integrate the individual to society, democracy to integrate social organizations to the individual.
>
> There is yet no ethic dealing with man's relationship to land and non-human animals, and plants, which grow upon it. Land, like Odysseus's slave girls, is still property. The land relation is still strictly economic, entailing privileges but not obligations.[2]

Leopold then cites the fact that in many places desecration of certain locations is strongly condemned and cites "Ezekiel and Isaiah have asserted that the despoliation of land is not only inexpedient, but wrong"; hence

[1] Leopold, Aldo. 1933, "The Conservation Ethic", *Journal of Forestry*, 31 6.
[2] ibid. p. 2.

he regards the "conservation movement as the embryo of this belief".[3] In fact, as we shall see in Chap. 5, these words of an inspired forester will eventually be re-echoed, mutatis mutandis, by those of a famous judge many decades later: J. Christopher Weeramantry.

Starting from this early beginning, it would be both impossible and counterproductive to cite and discuss all the scientific and related material that has addressed the issues of our concern. So, I will make no apologies for using primarily and almost exclusively material consonant with ecological integrity, often, but not exclusively, arising from the work of a group I have directed for twenty-seven years, and which numbers several hundred scholars, scientists, ethicists and primarily now legal scholars. I have started and directed the Global Ecological Integrity Group starting in 1992, and some of its history is indicated in the website.[4]

I believe that such an approach will be entirely appropriate since I will argue and then conclude that the direct line of thought starting precisely from Aldo Leopold's work to my own position on ecological integrity would have been the best road to allow in order to prevent the present impasse regarding both climate and food. Of course, many others have appreciated the importance of land and forests, beyond those who give primacy to integrity, and the independent conclusions they have reached, primarily regarding food and agriculture in the first part of this chapter, tended to complement and support our own work.

Particularly important is the work of Wendell Berry,[5] and we will discuss especially his Chapter 7, titled "Agricultural Solutions for Agricultural Problems dating from 1978" (pp. 113–124). Berry argues that the "most destructive change in modern times" has been the metaphor of "the machine":

> Until the industrial revolution occurred in the minds of most people in the so-called "developed" countries, the dominant images were organic: they had to do with living things, they were biological, pastoral, agricultural or familial.[6]

The changed vision now emphasizes efficiency. Agriculture is now viewed as agribusiness, the farm as a factory and plants, animals and the land itself

[3] Ibid., pp. 2–3.
[4] www.globalecointegrity.org
[5] Berry, Wendell, 1981, The Gift of Good Land, North Point Press, St. Francisco.
[6] ibid., p. 113.

as interchangeable parts or "units of production". This vision manifests the acceptance of industrial agriculture, which gives rise to several problems:

> (l) soil erosion, (2) soil compaction, (3) soil and water pollution, (4) pests and diseases resulting from monoculture and ecological deterioration.[7]

Further Berry also sees one connection to what we see today as one of the bases of climate change. He says:

> The energy industry, for instance, is now a cycle, but only a short arc between an empty hole and poisoned air.[8]

Farming, he argues, instead, is "inherently cyclical, capable of regenerating and reproducing itself indefinitely", something that echoes precisely the scientific sources that informed the original understanding of ecological integrity.[9]

In contrast, industrial agriculture treats soil as a "raw material which then it is free to use". Berry adds:

> ...at the present rate of cropland erosion Iowa's soil will be exhausted by the year 2050.[10]

Agribusiness's solution is to add chemical fertilizers, which renders the soil itself and everything grown on it toxic, instead of healthy and desirable. What is at stake is biodiversity, primarily in defense of wildlife, and its protection, which is basic to the respect for land.[11] Any defense of land, species and even forests, as we shall see, is based on the understanding of the foundational role of ecosystems, a role which will depend exclusively "on the scale at which one is approaching the natural world".[12]

[7] ibid., p. 115.

[8] ibid., p. 117.

[9] Hollings, C.S., 1992, "Cross-scale Morphology, Geometry, and Dynamics of Ecosystems", in *Ecological Monographs*, 62, No.4, pp. 447–502; see also Westra, L., 1994, *The Principle of Integrity*, Rowman Littlefield, Lanham, MD, pp. 43–51.

[10] ibid. p. 117.

[11] Noss, Reed P. and Cooperrider, Allen Y.,1994, *Saving Nature's Legacy*, Island Press, Washington, D.C.

[12] O'Neill, R.V., De Angel is, D.L., Waiide J.B. and Allen T.F.H., 1986, *A Hierarchical Concept of Ecosystems*, Princeton University Press, Princeton, N.J., p. 83.

When we turn to another perspective regarding conservation and species protection, we see the emphasis on fitness for survival, contrasted with the loss of genetic potential, which is introduced by any artificial intervention.[13]

What is significant is that these diverging, perhaps even contrasting, views can be easily reconciled under the perspective of ecological/biological integrity, which does encompass both. It seems as though we have veered away from both agriculture/starvation and climate change our main intended focus. But the important aspects of biodiversity we have described through different scientific approaches confirm the need to protect land from artificial intrusions which, one and all, are particularly significant for our topic.

For instance, the appropriately titled work of Herman Daly and John Cobb Jr., *For the Common Good*, clearly addresses this issue. The authors remark that "helpful government policies are possible":

> The most important would be taxing oil to include all externalities. The true price of oil-based agriculture would then become apparent, and monocultural production on huge acreages oriented to supplying distant markets would be at a disadvantage in relation to small farms raising a variety of foods for nearby consumers.[14]

It is easy to trace the importance of the premonitions here contained, which have been completely ignored, and eventually led to the current burning of the Amazon Forest, culminating as a global issue on the world stage in 2019, and we will discuss that issue in detail below. This will follow the argument of Daly and Cobb and further their previous collaboration with Robert Goodland and others.[15]

[13] Soule', Michael E., "Thresholds for Survival: Maintaining Fitness and Evolutionary Potential", in Soule', Michale and Bruce A. Wilcox, eds., 1980, Conservation Biology (An Evolutionary-Ecological Perspective), pp. 151–169.

[14] Daly, Herman E. and Cobb, John Jr.,1989, *For the Common Good*, Beacon Press, Boston, MA, p. 272.

[15] Goodland, Robert and H.E. Daly, 1998, "imperatives for environmental Sustainability Decrease Overconsumption and Stabilize Population" in *Population and Global Security*, N. Polorenin ed. Cambridge University Press, N.Y., pp. 117–132; Robert Goodland and David Pimentel, 2000 "Environmental Sustainability and Integrity in the Agricultural Sector", in D. Pimentel, L. Westra and R.N Hoss. eds., *Ecological Integrity: Integrating Environment, Conservation and Health*, Island Press, Washington, D.C., pp. 121–137 (note both Goodland and Pimentel as well as Noss were members of the Global Ecological

3.1.1 *Sustainability and Overconsumption*

Thus, we are faced globally not only with the hunger of the poor, who are deprived of any food whatsoever, but also with the Hunger of those who live in affluent countries. They have plenty of available food, but the way it is grown and then promoted renders it unsafe to eat, hence unable to safely satisfy hunger: I used the diction Hunger to indicate what most of the Western world is unfortunately suffering from.

This situation, I argued, is especially grave in North America, but even Europe, especially in its northern nations, is increasingly prey to the dangerous corporate ploys intended, as Pimentel and Goodland argued, to enrich some corporations, while attacking the very life and safety of citizens who have little knowledge and even less choice to avoid the major problems: the presence of herbicides and pesticides, primarily those manufactured by Monsanto/now Bayer, as research increasingly showed the toxic effects of genetically modified organisms (GMOs) and related pesticides and weed killers in their products, such as Roundup.[16]

We will return to the topic of corporate power below; for now it is important to isolate the major culprit, that is, Monsanto's glyphosate, the main ingredient of Roundup pesticide, together with the main product of that corporation, genetically modified organisms (GMOs), which in turn facilitate its spread. Also, there are at least two more important aspects of the problematic function of agriculture, primarily in the American continent: the first is the use of agriculture for the production of ethanol for energy.[17] Pimentel says, "using corn, a basic human food resource for ethanol production raises ethical and moral issues":

> At present world agricultural land supplies more than 99.7 percent of all world food (in term of calories) while aquatic systems supply less than 0.3

Integrity Group, Pimentel as a soil scientist at Cornell, Goodland working with Indigenous peoples' rights at the World Bank).

[16] Westra, L. 2018, *On Hunger: Lou v. J.*, J., 2013; *Monsanto v. The World*, Ultraculture Press, pp. 7–8; Stehle, S. and Schulz R., 2015, "Pesticide authorisation in the EU-Environment Unprotected?" *Environmental Science and Pollution Research* 22(24): 19, 19, 632–19, 647.

[17] Pimentel, David, 2009, "The Ecological and Energy Integrity of Corn Ethanol Production", in L. Westra, K. Bosselmann. and R. Westra eds., *Reconciling Human Existence with Ecological Integrity* Earthscan, London, pp. 245–255.

percent per capita, available cropland decreased by 20 percent, irrigation land by 1.2 percent.[18]

The other aspect of the problem, which is even more relevant given the focus of the present work, is the Report of Olivier de Schutter, which we cited and discussed in detail in the previous chapter. It is worth repeating:

> The FAO Study estimated that the livestock sector was responsible for 18 percent of greenhouse gas emissions measured in CO_2 equivalent, a larger share than transport. Once livestock respiration and the loss of greenhouse gas reductions from photosynthesis that are foregone by using large areas of land for grazing or feedcrops are taken into account, livestock is found to be responsible for 51 percent of anthropogenic gas emissions.[19]

Both these problems have been known, and that intelligence has been increasingly disseminated and accepted. That, however, is not the case for Monsanto's criminal activities, that is, for the uncontrolled spread of GMOs and glyphosate, the main issue that has suggested the use of Hunger as rich/Western deprivation of nourishing, safe food. What is particularly significant is that the spread of that knowledge has been accelerating exponentially in a relatively short period.[20]

This acceleration has also been experienced by this author.[21]

The main reason for the proliferation of anti-Monsanto/Bayer information was clearly the many successful court cases against Roundup/glyphosate (if not yet GMOs themselves), that is, against Bayer in 2018–2019 for the proven cancer incurred by those who had been in

[18] ibid. p. 251.

[19] De Schutter (2014).

[20] Gillam, Carey, 2017, *Whitewash* (The Story of a Weedkiller, cancer and the Corruption of Science), Island Press, Washington, D.C.

[21] Note: in 2017 *On Hunger* was accepted by Edward Elgar, London, and put under contract. In March 2018, after I saw my first proofs, I received a letter from EE's lawyers, expressing their desolation, apologies, and so on, but, after their copy editor had read my work, their fear of corporate reprisal was such that they would not be able to publish my work. It then took me several months to find a minor US press in California, which expressed no concern, and the work was published late in 2018. In contrast, in the Spring of 2019, I received a funded invitation by Erasmus University in Rotterdam, The Netherlands, to present work in a workshop devoted to glyphosate, which involved the WHO, IARC, Collegium Ramazzini and numerous government officials from the EU, as well as the Netherlands, where EFSA and ECHA ware explicitly attacked, as was Bayer, a silent presence at that meeting.

contact with that substance, and the website reports, almost too numerous to cite that culminated in September 2019, when it was reported that Germany joined a number of EU countries that explicitly banned glyphosate by the end of 2023.[22] In fact, the dangers of glyphosate and the correspondingly huge losses faced by Bayer's own stock after having to disburse millions to compensate users even convinced Walmart to discontinue the sale of Roundup.

Yet, the acceptance of the connection between glyphosate and cancer, now well-proven and accepted, only represents the tip of the iceberg regarding the toxicity of glyphosate and the corresponding criminality of those who manufacture it and the complicity of EFSA, the so-called European Food Safety Authority. The Ramazzini Institute published further chilling results: glyphosate is also an endocrine disruptor and the source of all the grave harms caused by hormonal imbalances and well documented for many decades by the WHO.[23]

These results are supported and in fact anticipated by two brave works published by Oxford University Press.[24] I say "brave" because although neither work was actually taken seriously by policymakers and government officials, when properly understood, these publications brought into question two of the topics considered taboo today: the almost unlimited rights of corporations and the never discussed, unscientific and never truly argued feminist mantra that a child is only such when it comes out of the womb.

At any rate, the connection between respect for the Earth and its creatures and the grave human harms we have alluded to is undeniable.[25]

[22] Agence France-Press, 4 September 2019, https://www.theguardian.com/environment/2019/sep/04/germany-ban-glyphosate-weedkillerby-2023

[23] Dr. Betty Martini, Founder Mission Possible World Health International, http://www.mpwhi.com, www.no.net, "New Results. on Glyphosate Dangers from the Ramazzini InstituteHormonal Imbalance, More", https://rense.com/general96/new-results-on-glyphesate-danger-from-the-ramazzini-institute.php, viewed 22/4,19.

[24] Landrigan, Philip and Etzel Ruth, 2013, *Textbook Children's Environmental Health*, Oxford University Press, New York; Grandjean, Philippe, 2013, *Only One Chance (To Develop a Brain)* Oxford University Press, New York.

[25] Kendra Klein and Anna Lappe', Wd., 7 August 2019, "American Agriculture is 48 times more toxic than 25 years ago. Blame Neonics", https://www.theguardian.com/common/is free/2019/Aug/07/Americas-dependence-on-pesticides-especially-neonics-is-a-war-on-nature" viewed Aug. 8, 2019.

3.1.2 Forests and the Life Within Them

The close connection between land/agriculture and land/forests is simple: it is the life within both. Orie Loucks considers how much wild land, or ecointegrity, we need for forests' health. His discussion is rife with complex calculations, but his conclusion is both simple and startling:

> When causal agents for the impairment of integrity are present with the potential to induce still more loss, a pattern of ecosystem destabilization can develop that leaves the system largely without useful functions[26]

Forests, like agricultural lands, nourish and maintain many creatures, beyond what they might indirectly do for human—and the biomass one may find in the land where trees grow include the lowly, but absolutely necessary earthworm.[27]

When presenting his findings at one of our meetings, Loucks explained that the loss or earthworms was a particularly grave issue, since they are the decomposers, necessary to ensure the elimination of dead animals. Thus, long before concerns emerged regarding the megafauna of the forests, the humble earthworm needed land and forests uncontaminated by toxic substances to survive.

When natural systems are not under attack, the life within them is both healthy and abundant. This is clearly in evidence when we move from terrestrial insects to those who provide nourishment for fish in running waters, and this when studied provided the basis for biological integrity and the corresponding index, created by James Karr, the Index of Biotic Integrity (IBI), following upon his research in the rivers where salmon lived and traveled in the US state near the University of Seattle.[28] As we consider the role of salmon in forest rivers, we also need to acknowledge that food is more than nourishment: for most of the Indigenous peoples,

[26] Orie Loucks, 2000, "Pattern of Forest Integrity in the Eastern United States and Canada: Measuring Loss and Recovery, in D. Pimentel, L. Westra and R. Noss eds, *Ecological Integrity: Integrating Environment, Conservation and Health*, Island Press, Washington, DC, pp. 177–190 p. 188.

[27] Olson, J.S., 1963). "Energy Storage and the Balance of Producers and Decomposers in Ecological Systems", in *Ecology*, 44(2): 322–331; Reichle, D.E., 1977, "The Role of Soil Invertebrates in nutrient cycling" in V. Lohm and T. Persson eds., *Soil Organisms as Components of Ecosystems*, Ecological Bulletins, 25: 145–156.

[28] Karr, James R., 2000, "Health, Integrity and Biological Assessment: The Importance of Measuring Whole Things", in Pimentel, Westra and Noss, op.cit. pp. 209–226.

food holds a deep, traditional, often religious significance. That is particularly true of the First Nations of Canada.

Salmon and moose are more than food and are in fact protected by Canadian law as the right for all to practice their religious traditions and maintain their cultural integrity. Shaun Fluker explores the protection of ecological integrity in the natural forest areas such as the Wood Buffalo National Park in the Canadian province of Alberta, extending all the way to the Northwest Territories.[29] He shows how, although these Canadian forests are protected and in fact all decisions which affect them should be governed by the primacy of ecological integrity, as stated in the national parks regulations, essentially the final decision in that case was based on a different law. The laws governing the First Nations rights to hunting and fishing are supported by the Canadian Constitution.[30]

The issue was a commercial enterprise's request to construct a highway through the forest for easier access to points north. But whenever a highway passes through a wild forest, it disrupts the wildlife within it, as each species needs to be able to reach others of its own kind with whom to mate, without impediment or the species to survive. Hence the integrity of the forest, although it was at risk, was eventually not protected as such, but as the hunting and fishing constitutional rights of the local bands of First Nations. We will return to the question of Indigenous peoples in general in Chap. 5.

The main concern now is the function of the forests themselves and that of the wildlife they support. The problem is that while healthy land is the only assured way to promote safe agriculture, hence food, forests are also basic for the protection of a stable climate; to use a phrase present in all 2019 news programs, whether on television or newspapers, forests are now said to be "the lungs of the world", a reality ignored for far too long.

At the time of early environmental concerns, forests were protected in order to protect/conserve the wildlife, the megafauna within them. For instance, that was the approach of Reed Noss, particularly of his work in

[29] Fluker, Shaun, 2013, "Environmental Norms in the Courtroom: The Case of Ecological Integrity in Canada's National Parts", in L. Westra, P. Taylor and A. Michelot, eds, *Confronting Ecological and Economic Collapse*, Earthscan Routledge, London, UK, pp. 21–35.

[30] Constitution Act 1982, Being Schedule B to the Canada Act, 1982 (UK)c.11. Section 35(1) reads: "The Existing aboriginal and treaty rights of the aboriginal people of Canada are hereby recognized and affirmed".

the "Wildlands Project".[31] In 1992 climate was not a major issue, and neither was hunger. Conservation of forests was a significant issue in order to protect natural habitats and, as mentioned, the mating habits of the animals within them.

The "Wildlands Project" envisions protecting forests from those in the US West all the way to Alaska: a guaranteed protected corridor where wildlife would roam undisturbed by commercially motivated enterprises. Of course, such protection would also guarantee the ecological integrity of the forests which, as we saw, was at least formally a requirement according to the regulations of the Canadian National Parks. These regulations, however, were not present in US legal regimes nor in countries south of the US, in the South American continent, where US-based corporations enjoy total impunity for their operations intended to enrich their owners and shareholders, as well as enhance the political power of the US.

None of the commercial enterprises either in the US or in countries under US influence were governed by any regime intended to protect the forests, the wildlife or even the humane that made their living within them. Biologists have been sounding the alarm for years regarding the depletion of species, as the numbers grew increasingly alarming, while the proposed attempts to reverse those trends remained unsuccessful today.

In fact, we are reaching an "extinction crisis", as we are living in the sixth wave of extinctions in the past half-billion years: as many as 30–50 percent of all species are heading to extinction by 2050. It is a proven fact that 99 percent of currently threatened species are in their predicament as a result of human activities, including losses caused by climate change. Species diversity ensures ecosystem resilience, and its current trend represents a dangerous ongoing loss of habitat integrity. Losses include amphibians, birds, fish and invertebrates as we saw. Mammals are among the most affected extinct or at grave risk, with 5491 known mammals declining in population, and 68 percent of the plants that are the foundation of life on Earth are also extinct or threatened with extinction.[32]

This topic goes well beyond the aim of our present work; thus, we should return to today's forests which exhibit the results of all the neglectful, uncaring and short-sighted decisions and practices that have affected

[31] Noss, R., 1992, "The Wildlands Project, L Land Conservation Strategy", in Wild Earth, Special Issue, pp. 10–25.

[32] Centre for Biological Diversity, http://www.biologicaldiveraity.org/programs/biodiversity/elements_of_biodiversity/extinction_crisis/, viewed Sept. 25. 2019.

them. As suggested in the discussion on hunger, however, the final, unavoidable awareness of the interface between the destruction of forests and climate change was long denied by those who have been termed "merchants of doubt".[33] Most today are familiar with the long saga of denials that permitted and in fact fostered the death of millions of smokers, until finally the WHO/WHA introduced the Tobacco Convention, which put period to the lies, the doctored "facts", the corruption of science, which has now shifted to food.[34]

The authors, however, primarily focus on global warming, as they trace the history of the development of denial of climate change.[35] Hence the stage was set long ago for the present dire situation, which at this time is only described in the details available on websites, as the awareness of the issue seems to have peaked in the year 2019. We will discuss the main issues in the sources here listed.[36]

Perhaps we can start with a scary overview by Danielle Garrand for CBS News.[37] Sao Paulo, Brazil, went dark in the afternoon because of the multiple fires set and burning in the Amazon, and Reuters reported the black smoke blanketing Sao Paulo which was reported by NASA as visible from outer space. It affected not only the regions of Amazonia, Mato Grosso and Rondonia but also neighboring states Bolivia and Paraguay.

[33] Oreskes, Naomi and Conway, Erik M., 2010, *Merchant of Doubt*, Bloomsbury Press, New York.

[34] The WHO FCTC opened for signature on June 16 to June 22, 2000, in Geneva and thereafter, at the United Nations Headquarters in New York, the Depository of the Treaty, from June 30, 2003, to June 29, 2004. The Treaty which is now closed for signatures has 166 signatories, including the European Community. The Convention entered into force on February 27, 2005—90 days after it had been acceded to, ratified, accepted or approved by 40 states.

[35] Oreskes and Conway (2010: pp. 169–215).

[36] Fanf Lee, Aug. 23, 2019, "GOP Lobbyists Help Brazil Recruit US Companies to Exploit the Amazon", viewed Aug. 26, 2019 at https://www.theinteroept.com/2019/08/2J/gop-lobbyistshelp-brazi1-recruit-u-s-companies-to-exploit-the-amazon/; Zaitchik, Alexander, July 6, 2019, "Rainforest on Fire-On the Front Lines of Bolsonaro's War on the Amazon, Brazil's Forest Communities Fight Against Climate Catastrophe", in partnership with the Pulitzer Centre, viewed on Aug. 26, 2019 at https://theintercept.com/2019/07/06/brazil-amazon-rainforest-indigenoua-conaervation-agribusiness-ranching

[37] Garrand, Danielle, Aug. 21. 2019, "Parts of the Amazon rainforest are on fire, and smoke can be spotted from space", viewed Aug. 22, 2019 at https://www.cbs.news.com/news/amazon-wildfire-parts-ot-amazonrainforest-on-fire-smoke-seen-from-space-2019-08-20/

This was not an occasional fire of course; it was part of a carefully orchestrated campaign to clear the forest for business and for the interest of Brazil's Bolsonaro. Lee Fang describes the active role played by the US Republican party in aiding Jair Bolsonaro by recruiting commercial enterprises to further investments in the Amazon, which entailed cutting down the forest and setting fires to facilitate the plans to produce fodder for industrial agriculture and even energy substitutes, rather than any kind or food. Trump himself supported Bolsonaro's victory, rising from obscurity in his country, to be known as "Captain Chainsaw" for his role in demolishing the forest. Rudy Giuliani, Trump's legal associate, also invested over one and a half million in those projects.

We will return to the corporate issues behind the US involvement in Chap. 4. For now Alexander Zaitchik, in partnership with the Pulitzer Centre, describes the grave human suffering that follows when the "rainforest is on fire"; Bolsonaro's own practices are described as "war on the Amazon", given the effects his policies have on the inhabitants of that forest. The huge area of the Amazon is covered by

> trees and vegetation so dense and heavy with water, that it exhales a fifth of the Earth's oxygen, stores centuries or carbon, and deflects and consumes and unknown but significant of solar heat. Twenty percent of the world's freshwater cycles through its rivers, plants, soil and air. This moisture fuels and regulates multiple planet-scale systems, including the production of 'rivers in the air' by evapotranspiration, a ceaseless cleansing flux in which the forest breathes its water into great hemispheric conveyor belts that carry it as far as the breadbaskets of Argentina and the American Midwest, where it is released as rain. (pp. 1–2)

This long paragraph captures well the inescapable connection between climate and food production, and it emphasizes the unique role of this forest, just as the previous article emphasized the criminality of activities based exclusively on the economic interests of non-state actors, as well as the connection between those MNEs and a specific country, the US.

This report is particularly rich and complete regarding human rights abuses, and we will cover that aspect of the issue in Chap. 4. Now the most important point is the connection between climate change and commercial policies, which ignore the importance of those links, those responsibilities and the worldwide implications they carry. The World Wildlife Fund addresses the joint problem of deforestation and forest degradation as well.[38] Forests cover 31 percent of the planet's land. Their role is to purify water and air, and "eighty percent of the world's land-based species live in forests".

The Amazon is the clearest case in point as almost 20 percent of it has already been lost to deforestation. But Sumatra, Indonesia, is also at grave risk, as the forest is cleared to make way for palm oil plantations, and it is becoming another one of the spots at grave risk.

> Up to 420 million acres of forest could be lost between 2010 and 2030 in these "deforestation fronts" if current trends continue. The hot spots are located in The Amazon, the Atlantic forest, and Gran Chaco, Borneo, The Cerrado, Choco-Darien, The Congo Basin, East Africa, Eastern Australia, Greater Mekong, New Guinea and Sumatra. (p. 3)

In all these cases, the attacks on human rights, the basic racism, are rampant, as the forest's habitat becomes increasingly dangerous to the wildlife within it, and with the presence of wildlife and their interactions, the ecological integrity of the forests is also lost.

We can conclude this chapter with yet another discussion of the Amazon, as Ryan Grim returns to another aspect of its exploitation.[39] Grim speaks of a water terminal in Amazon run by Hidrovias do Brazil (owned primarily by Blackstone). The aim was the exportation of soybeans, intended to reach the port through a road, originally unpaved but improved since 2016. The corporation was keen to "develop" the road, in order to truck soybeans (now entirely genetically modified and pesticide resistant), from the state of Mato Grosso.

But modern highways contribute significantly to deforestation, which was also supported by the right-wing government under Blairo Maggi, a soy mogul (more details are given in Chap. 5).

Bolsonaro wanted the Amazon turned over to industrial agriculture; hence the forests were viewed simply as an impediment to his plans. Thus, while the GMO soybeans were disseminated in the world, rendering food unsafe in developed countries, the forest's role and function were undermined with no consideration for their foundational role for the climate

[38] World Wildlife Fund, "Deforestation and Forest Degradation", viewed Sept. 23, 2019 at https://www.worldwildllfe.org/threats/deforestation-and-forest-degradation

[39] Ryan Grim, August 27, 2019, "A Top Financier of Trump and McConnell is a driving force behind Amazon deforestation", viewed Sept. 21, 2019 at https://theintercept.com/2019/08/27/amazon-rainforest-fire-blackstone/

and for their own biodiversity. In the next part, we will discuss briefly the role of water and air, as they affect both climate and food production and availability.

3.2 Part 2: The Role of Water and Air

3.2.1 Water's Nature and Role

> The most noteworthy characteristic of any small body of fresh water – be it a pond, stream, an icicle or a rain cloud – is impermanence...In a word it is dynamic. In much the same way as that, every living organism has a life cycle, water has a life cycle: it circulates. Indeed, all the water on earth is constantly circulating.[40]

That nature indicates that it is indeed appropriate that water has been used to establish the deep significance or biological integrity, it, too, is "dynamic" in the sense that organisms are. At any rate, the role of water is deeply significant for our chosen topic, in relation to starvation in two separate but related aspects. When water is no longer present because of extreme high temperature, as it was in the northern African regions, desertification occurs: the land cracks and breaks apart and will no longer produce any food; hence it shows a clear deprivation of human rights in the area.

Also, as the chemical/toxic inputs we described as related to agriculture in the previous chapter leach into the waters, even if an area might still produce food, it is no longer nutritious or safe, so that Hunger occurs, as previously explained. Water is the substance of life: we live in it before emerging into the world, but water, just as earth hosts incredible numbers of wildlife, hosts themselves of microscopic creatures:

> The organisms are too small to be examined with the unaided eye, even those that are just visible as tiny specks, but they are abundant enough to affect the water in which they live, in a way that influences the welfare of all the larger animals and plants sharing it.[41]

Some of these tiny creatures are bacteria, and "without bacteria altogether life on earth would soon grind to a halt...Before plants can absorb

[40] Pielou, E.C., 1998, *Fresh Water*, University of Chicago Press, Chicago, p. 1.
[41] Pielou (1998: 227).

nitrogen, it must be "fixed", that is, converted into simple inorganic compounds that plants can absorb, and nearly all the fixing is done by bacteria".[42] Thus the water is so vital for the functioning of the climate, but it is at the same time also basic to the normal functioning of the earth and its capacity to provide nourishment for both animals and humans.

We discussed the importance of keeping water unpolluted as only then the organisms that live within it will be able to sustain their necessary functions, hence the necessity for what the 1972 US Clean Water Act demands: the protection of the water's biological integrity. The Great Lakes Water Quality Agreement emphasized that message as at least one of the Great Lakes of Canada, Lake Erie, was in fact pronounced dead, that is, it was so choked by algae and foreign substances that it could no longer support life, such as fish.[43]

Of course, the lack of fish in a lake, at first sight, does not appear to be a situation leading to starvation, but, especially in the Great Lakes region, inhabited as it is by many First Nation communities, the situation is indeed grave (see Chap. 4). When water is polluted, disaster may follow, and when water is totally absent, starvation ensues. Thus, extreme heat causes desertification, and desertification causes starvation, which in turn causes also the flight of desperate people from affected areas, as climate refugees (Chap. 4).

Water is also the main issue as we consider, once again, the burning forest in the Amazon. The problem is one of the necessary evaporation, which is part of the water cycle:

> Vegetation pumps an enormous amount of water from the soil into the air, few people realize how much, because the whole process is invisible. For example, a single hectare of Douglas-fir forest spews out about 59 tons of water vapor in the course of a sunny summer day, or about 235 bathtubs full. Evaporation from the soil and transportation from the vegetation combined is known as evapotranspiration.[44]

[42] ibid. p. 234.

[43] Regier, H. et al. "Integrity and Surprise in the Great Lakes Basin Ecosystem" *in An Ecosystem Approach to the Integrity of the Great Lakes in Turbulent Times*, Great Lakes Special Publication 90–4, Ann Arbor, Mich., Great Lakes Commission, 1990: 17–36; Barlow, Maude and Clarke, Tony, 2003, *Blue Gold* (The Battle Against Corporate Theft of the World's Water), "Losing the Great Lakes" pp. 35–6, McLelland and Stewart, Toronto, Ont.

[44] Pielou, op. cit, p. 240.

Hence, the enormous damage deliberately burning a forest for other economic purposes does to the natural functioning of water in relation to climate must be viewed as a crime against humanity as a whole.

3.2.2 The Legal Position of Water and Water Rights

> While it is one of the most essential resources for human survival, the quantity of fresh water on the planet is essentially fixed or declining…The result is considerable stress on existing legal regimes as they attempt to respond to increasing and changing demands for water without unduly destabilizing existing expectations expressed in restructures in water using facilities.[45]

In fact, the WHO/UNICEF (2002) showed that over 1.2 billion people do not have access to safe drinking water and 239 billion people lack proper sanitation.[46] Also, water has often been transformed into private property or public property; thus it seems urgent to go beyond establishing these and other facts and to institute instead a legal right to water, although,

> Thus far, no treaty or other binding legal document expressly recognizes a human right to water, although arguably, such a right is inferable from a number of human rights treaties and other legal documents.[47]

Accordingly there is a clear enough legal basis to operationalize the human right to water and the obligation of states to respect that right and to take steps to protect it, and that right has been also clearly expressed in Article 17, The Right of Access to Water, of the International Law Association.[48] Joseph Dellapenna is a legal scholar, and he outlines clearly what should be morally and legally right. However, we have been noting what is actually happening in reality, and we will return to that reality here, before returning to the main reasons for that divergence in Chap. 5.

[45] Dellapenna, J, 2008, NA Human Right to Water" in L, Westra, K. Bosselmann and R. Westra eds., *Reconciling Human Existence with Ecological Integrity*, pp. 183–194, 184.

[46] WHO/UNICEF, 2002, "Global Water Supply and Sanitation Report" available at www.who.int/water_health/globalasseasment/GlobalTOC.htm, p. 185.

[47] ibid., p. 286; see UNECOSOS, General Comment 15 which expressly recognized the existence of a human right to water based upon several articles of the International Covenant.

[48] International Law Association, 2004, "Berlin Rules on Water Resources" in *Report or the Seventy-first Conference, Berlin*, 2004, International Law Association, London.

The lack of clean, safe, healthy water has created yet another water issue: the need to rely on bottled water of various kinds, now ubiquitous everywhere. From gyms to schools and college classrooms, from offices to homes, those who can afford it replace the potentially unsafe tap water with bottled water. In turn this creates yet another grave water-related problem, that is, the presence or huge amounts of plastic in our oceans: a threat to all life there, as well as to the integrity of ocean water.[49] This is yet another example of a technological fix that adds another layer of difficulty and harm, as the choice was purely based on the promise of economic profit.

Many writers also note the problem of unequal access to water: as with food, there ought to be access to safe, clean water, but that is not the case in too many countries:

> There is nowhere on earth to go to escape the global water crisis. For instance, in wealthy United States, the Natural Resources Defense Council says that some 53 million Americans, nearly one-fifth of the population drink tap water contaminated with lead, fecal bacteria, or other serious pollutant.[50]

We noted the same problem in the case of food deprivation: Hunger described the lack of safe, healthy food in wealthy developed countries. Barlow and Clarke, of course, also acknowledge that the third world is deprived of water altogether, and we will return to that issue in Chap. 4. Hence water is biologically necessary to stabilize the climate, to nourish the earth and to provide food. It is also vitally necessary for human survival; although the existing legal instruments should be used to cement both moral and legal rights in principle, in reality the right to water is far from secure. Air is the final element; we need to discuss its physical presence and its role.

3.2.3 *The Role of Air*

The dangerous role of air in climate change is obvious: tsunamis, hurricanes and all extreme weather events are accompanied by high winds. Even floods, hail and other such climate-induced problems occur with

[49] Barlow and Clarke, op.cit. pp. 143–145.

[50] Barlow, Maude and Clarke, Tony 2002, *Blue Gold* (The Battle against Corporate Theft of the World's Water), McClelland and Stewart Ltd., Toronto Ontario, p. 58.

heavy winds. What of the air itself? At least, thus far, there has been no clear effort to commercialize it, privatize it and sell it back to us at a profit. And yet, clean air is no more available to all than clean water is, both in wealthy countries and in poorer ones.

One need only consider the plight of the Chinese peoples: most group/community pictures in the media show them wearing masks, in order to avoid the smog and other harmful air conditions. We also saw the thick black smoke over Sao Paulo, Brazil, obviously harmful, although those affected were not protected in any way and probably could not afford masks. It seemed like a minor side effect of the burning Amazon Forest, but—no doubt—the same dangerous smoke conditions were present all over Amazonia, yet another hazard to be faced by the area's native populations.

In wealthy countries, as smog prevails in many large cities, automobile traffic may be limited in certain days, and even in those cases, we see people forced to wear masks, while rates of asthma and other respiratory diseases grow exponentially, particularly in larger urban centers. Nevertheless, there is no legal treaty, declaration or other legal instruments to guarantee our right to air or to safe breathing. It is a sad situation indeed, to be forced to acknowledge that even our right to breathe safely ought to be protected by legislation, as we can no longer take such right for granted. As long as air is polluted, imbued with toxic, chemical and other foreign substances, its integrity is also no longer there, and we shall return to this point in our final chapter.

We need land, forests and water, but we also need air and should be entitled to it, and as with food and water, it should be readily available, safe and healthy. It is not either for the wealthy or for the poor. In relation to climate change, just as water warming is increasingly causing harms and may even kill the wildlife within it, so too the heavy, high humidity air; when temperatures soar beyond 40 degrees just as have been observed primarily in northern Europe and in Canada and the US since 2018–2019, the result is harm to many animals and to humans; the air then becomes almost a weapon of mass destruction in most countries, and its safety should be legally protected as an integral harmful aspect of climate change results from its extreme heat, and in 2019 the children and young people of the world have been steadily proclaiming.

References

Agence France-Press, 4 September 2019, https://www.theguardian.com/environment/2019/sep/04/germany-ban-glyphosate-weedkillerby-2023

Barlow, Maude and Clarke, Tony 2002, *Blue Gold* (The Battle against Corporate Theft of the World's Water), McClelland and Stewart Ltd., Toronto Ontario, p.58

Berry, Wendell, 1981, The Gift of Good Land, North Point Press, St. Francisco

Centre for Biological Diversity, http://www.biologicaldiveraity.org/programs/biodiversity/elements_of_biodiversity/extinction_crisis/, viewed Sept. 25. 2019

Constitution Act 1982, Being Schedule B to the Canada Act, 1982 (UK)c.11. Section 35(1) reads: "The Existing aboriginal and treaty rights of the aboriginal people of Canada are hereby recognized and affirmed"

Daly, Herman E. and Cobb, John Jr., 1989, *For the Common Good*, Beacon Press, Boston, MA, p.272

de Schutter, O., 2014, *Final Report: The Transformative Potential of the Right to Food, UN doc.GE.14–10537*, Human Rights Council 25th Session

Dellapenna, J, 2008, NA Human Right to Water" in L. Westra, K. Bosselmann and R. Westra eds., *Reconciling Human Existence with Ecological Integrity*, pp.183–194, 184

Fluker, Shaun, 2013, "Environmental Norms in the Courtroom: The Case of Ecological Integrity in Canada's National Parts", in L. Westra, P. Taylor and A. Michelot, eds, *Confronting Ecological and Economic Collapse*, Earthscan Routledge, London, UK, pp.21–35

Garrand, Danielle, Aug. 21. 2019, "Parts of the Amazon rainforest are on fire, and smoke can be spotted from space", viewed Aug.22, 2019 at https://www.cbs.news.com/news/amazon-wildfire-parts-ot-amazonrainforest-on-fire-smoke-seen-from-space-2019-08-20/

Gillam, Carey, 2017, *Whitewash* (The Story of a Weedkiller, cancer and the Corruption of Science), Island Press, Washington, D.C.

Goodland, Robert and David Pimentel, 2000 "Environmental Sustainability and Integrity in the Agricultural Sector", in D. Pimentel, L. Westra and R.N Hoss. eds., *Ecological Integrity: Integrating Environment, Conservation and Health*, Island Press, Washington, D.C., pp.121–137

Goodland, Robert and H.E. Daly, 1998, "imperatives for environmental Sustainability Decrease Overconsumption and Stabilize Population" in *Population and Global Security*, N. Polorenin ed. Cambridge University Press, N.Y., pp.117–132

Grandjean, Philippe, 2013, *Only One Chance* (To Develop a Brain) Oxford University Press, New York

Grim, Ryan, August 27, 2019, “A Top Financier of Trump and McConnell is a driving force behind Amazon deforestation”, viewed Sept. 2 1 2019 at https://theintercept.com/2019/08/27/amazon-rainforest-fire-blackstone/

Hollings, C.S., 1992, “Cross-scale Morphology, Geometry, and Dynamics of Ecosystems”, in *Ecological Monographs*, 62, No.4, pp.447–502

International Law Association, 2004, “Berlin Rules on Water Resources” in *Report or the Seventy-first Conference, Berlin*, 2004, International Law Association, London

Karr, James R., 2000, “Health, Integrity and Biological Assessment: The Importance of Measuring Whole Things”, in Pimentel, Westra and Noss, op. cit. pp.209–226

Klein, Kendra and Anna Lappé, Wd., 7 August 2019, “American Agriculture is 48 times more toxic than 25 years ago. Blame Neonics”, https://www.theguardian.com/common/isfree/2019/Aug/07/Americas-dependence-on-pesticides-especially-neonics-is-a-war-on-nature viewed Aug. 8, 2019

Landrigan, Philip and Etzel Ruth, 2013, *Textbook Children's Environmental Health*, Oxford University Press, New York

Lee, Fang. Aug. 23, 2019, “GOP Lobbyists Help Brazil Recruit US Companies to Exploit the Amazon”, viewed Aug. 26, 2019 at https://www.theinteroept.com/2019/08/2J/gop-lobbyistshelp-brazi1-recruit-u-s-companies-to-exploit-the-amazon/

Leopold, Aldo. 1933, “The Conservation Ethic”, *Journal of Forestry*, 31 6

Loucks, Orie, 2000, “Pattern of Forest Integrity in the Eastern United States and Canada: Measuring Loss and Recovery”, in D. Pimentel, L. Westra and R. Noss eds, *Ecological Integrity: Integrating Environment, Conservation and Health*, Island Press, Washington, DC, pp.177–190 p. 188

Martini, Dr. Betty Founder Mission Possible World Health International, www.mpwhi.com, www.no.net, “New Results. on Glyphosate Dangers from the Ramazzini InstituteHormonal Imbalance, More”, https://rense.com/general96/new-results-onglyphesate-danger-from-the-ramazzini-institute.php, viewed Apr. 22, 2019

Monsanto v. The World, Ultraculture Press, pp.7–8

Noss, R., 1992, “The Wildlands Project, L Land Conservation Strategy”, in Wild Earth, Special Issue, pp. 10–25

Noss, Reed P. and Cooperrider, Allen Y., 1994, *Saving Nature's Legacy*, Island Press, Washington, D.C.

O'Neill, R.V., De Angelis, D.L., Waiide J.B. and Allen T.F.H., 1986, *A Hierarchical Concept of Ecosystems*, Princeton University Press, Princeton, N.J., p.83

Olson, J.S., 1963). “Energy Storage and the Balance of Producers and Decomposers in Ecological Systems”, in *Ecology*, 44(2): 322–331

Oreskes, Naomi and Conway, Erik M., 2010, *Merchant of Doubt*, Bloomsbury Press, New York

Pielou, E.C., 1998, *Fresh Water*, University of Chicago Press, Chicago, p.1

Pimentel, David, 2009, "The Ecological and Energy Integrity of Corn Ethanol Production", in L. Westra, K. Bosselmann. and R. Westra eds., *Reconciling Human Existence with Ecological Integrity* Earthscan, London, pp.245–255
Regier, H. et al. "Integrity and Surprise in the Great Lakes Basin Ecosystem" *in An Ecosystem Approach to the Integrity of the Great Lakes in Turbulent Times*, Great Lakes Special Publication 90–4, Ann Arbor, Mich., Great Lakes Commission, 1990: 17–36
Reichle, D.E., 1977, "The Role of Soil Invertebrates in nutrient cycling" in V. Lohm and T. Persson eds., *Soil Organisms as Components of Ecosystems*, Ecological Bulletins, 25: 145–156
Soulé, Michael E., "Thresholds for Survival: Maintaining Fitness and Evolutionary Potential", in Soule', Michale and Bruce A. Wilcox, eds., 1980, Conservation Biology (An Evolutionary-Ecological Perspective), pp. 151–169
Stehle, S. and Schulz R., 2015, "Pesticide authorisation in the EU – Environment Unprotected?" *Environmental Science and Pollution Research* 22(24): 19, 19, 632-19, 647
UNOCOSOC, General Comment 15 which expressly recognized the existence of a human right to water based upon several articles of the International. covenant
Westra, Laura, 1994, *The Principle of Integrity*, Rowman Littlefield, Lanham, MD, pp.43–51
Westra, Laura, 2018, *On Hunger: Lou v. J.*, J., 2013
WHO/UNICEF, 2002, "Global Water Supply and Sanitation Report" available at www.who.int/water_health/globalasseasment/GlobalTOC.htm, p.185
World Wildlife Fund, "Deforestation and Forest Degradation", viewed Sept. 23, 2019 at https://www.worldwildllfe.org/threats/deforestation-and-forest-degradation
Zaitchik, Alexander, July 6, 2019, "Rainforest on Fire-On the Front Lines of Bolsonaro's War on the Amazon, Brazil's Forest Communities Fight Against Climate Catastrophe", in partnership with the Pulitzer Centre, viewed on Aug.26, 2019 at https://theintercept.com/2019/07/06/brazil-amazon-rainforest-indigenous-conservation-agribusiness-ranching/

CHAPTER 4

Aggravating Circumstances Corporate Power v. Democracy and Land Grabs for Fodder and Fuel While the Amazon Burns

Abstract The author argues that, while scientific evidence has been mounting and becoming increasingly clear regarding both climate change and inappropriate diets, there has been another unchecked and contrasting growth, that of the power of corporate persons and associations. That power is supported by legal regimes far more supportive of corporate rights than of scientific realities. The start of this dissonance of power coincides with the perversion of democracy in the US, which has grown steadily since the first time the US Constitutional Court has emitted a landmark sentence stating that election funding was equivalent to "free speech" (1976). It is entirely appropriate to show the US issue because it is there that is where most of the dangerous corporations threatening not only democracy but life have their origins. These corporate giants impose their will through another practice, now acknowledged to be a crime against humanity: land grabbing. These "grabs" involve primarily Indigenous peoples and the poorest of local inhabitants in all continents. Land grabs masquerade as normal and legal business transactions, but contracts and fair business deals are not possible between groups of different languages, abilities and interests, without impartial legal representation, which is not provided.

Keywords Democracy • Corporate power • US Constitutional decisions • Land grabbing • Free speech

L. Westra, *Climate Change and Starvation*, Palgrave Insights into Apocalypse Economics,
https://doi.org/10.1007/978-3-030-42124-3_4

4.1 Introduction to Aggravating Circumstances

Buckley v. Valeo (424 U.S., 1(1976))

> Is a Constitutional Law Case on Campaign finance. A majority of Justices held that limits on election spending in the Federal Election Campaign Act of 1971&608 are unconstitutional.[1]

Recently Noam Chomsky, speaking on a question and answer program on Russia Today, stated that the already-precarious state of democracy in the US received its final blow with this constitutional case. Peter Venton recently contrasted democracy with oligarchy, and—sadly—his description fits precisely the situation Chomsky described.[2]

It is entirely appropriate to initiate our discussion of corporate power with the situation present in the US since most of the major multinationals that control the world's economy and more are based in the US.

As we saw at the beginning of Chap. 3, concern and respect for land, forests, animals and plants started as far back as Aldo Leopold (1933). Concern for water, however, has been growing more recently. We discussed the approach to the issue in Canada, a wealthy country rich in water.[3] And we shall see Vandana Shiva as she discusses the corporate control of water in an impoverished population, in a country where too many are suffering.[4] Nor are land and water the only parts of nature that are severely affected for both rich and poor: as we saw the climate itself is everywhere a risk, and while air, climate change are not themselves at risk, we are all at risk, or at least headed for a "climate apartheid".[5] Syria and other North African countries are so affected by desertification that the earth is literally scorched, dry and cracking, incapable of feeding the local populations. Nor is this an unusual ease: poor countries are suffering from climate change effects much more than the wealthy ones, as pointed out

[1] 421 US 1, 96S.ct. 612, 46 L.

[2] Venton, Peter, 2020, The Political Economy of Managing Without Growth, in press.

[3] Barlow and Clarke 2003, op.cit.

[4] Shiva, Vandana, 2002 *Water Wars* (Privatization, Pollution and Profit), Between the Lines, Toronto, Ontario.

[5] Dockrill, Peter, *Science Alert*, June 26, 2019, viewed Sept. 22, 2019 at https://www.businesainsider.com/climate-apartheid-united-nations-report2019-6?IR=T

by Philip Alston, UN Special Rapporteur on extreme poverty, who blames climate change for the situation.[6]

Yet the connection between starvation and climate change is established beyond doubt, although neither the UN nor individual states are either capable or willing to respond to it in a way that would address these issues in a meaningful way. There are two problems, and after consideration, the two can actually be combined into one. The first problem is the weakness and inefficiency of the UN as even major problems that should be handled by the UN are handled by the Security Council, which is politically controlled rather than independent as it should be, and international law itself has not developed in a way to render it able to cope with today's novel and grave issues.[7]

The second problem is the corporate control of most resources, as we have argued in the previous chapter. Vandana Shiva shows the devastating effects of such control in her country—an example of what the poor of the world have to bear.[8] However when we confront these two issues, it becomes increasingly obvious that they can be combined into one major problem, that is, the corporate control of both state decisions, due to their decisive input in the elections of candidates, on one hand, and the corporate stranglehold on the world's resources practiced by these corporate citizens, on the other.

That stranglehold is also much in evidence when we consider another crime against humanity: the phenomenon of land grabs.[9] These grabs involve primarily Indigenous peoples and the poor, starting with the grave situation in Amazonia, where corporate land grabs are clearly responsible for the climate disaster caused by the fires that are destroying the region.

The same issue, that is, land grabs, affects equally the Northern African states. The result of both these crimes will be discussed in Chap. 5 where

[6] United Nations Office of the High Commissioner on Human Rights, Statement by Prof. Philip Alston, UN Special Rapporteur on Extreme Poverty and Human Rights, Malaysia, 13–23 August 2019.

[7] Westra, Laura, 2019, *From Principles to Norms: The Development of International Law*, Nova Science Publishers, New York.

[8] Shiva, Vandana, 2002, *Water Wars* – Privatization, Pollution and Profit, Between the Lines, Toronto, Ont.

[9] Liberti, Stefano, 2014, *Land Grabbing: Journeys in New Colonialism*, Verso, London; Westra, Laura, 2018, *Environmental Disasters and Land Grabs as Crimes Against Humanity*, Nova Science Publishers, New York.

desperate people attempting to escape the violence of conflicts as well as starvation are forced to face further victimization at the hand of extreme right parties in the countries that ought to offer aid and support instead.

The land grabs masquerade as normal business transactions, which cannot legally take place between such unequal groups, and based, as they are, on incomplete understanding of the terms offered. These deals are the continuation—in some sense—of the environmental racism practiced in the US especially many years ago and ongoing today, particularly in some southern states. Corporations bent on minimizing expenses while maximizing profits at the expense of vulnerable people, be they African Americans or Hispanics, living in the poorer areas, are told that the decisions that affect them are purely based on economic choices and justifiable on that basis: they are increasingly hard to believe today as information becomes available.[10]

4.1.1 Corporate Power Against Democracy

Buckley v. Valeo[11] ensured that, as we saw in Chap. 1, democracy could not survive the identification of money with speech. There was no possibility of discussion, of debate, of the use of reason to make the case for one candidate or another, one platform or another. That has been the basis of democracy from the time of Aristotle.[12] Nothing was left but a corporate preference, the candidate that would further the economic advancement of corporations and their power: right or wrong no longer had any place in political choice. All a US government had to do was and is to find a few sympathetic judges, elected not for their qualifications, but for their political leanings, and use them to rubber stamp his/their interpretation of the sacred US Constitution.

We see the results almost daily in the news, as one illegal shooting follows another, at least once a week if not almost daily, in school, churches, mosques, shopping malls, as high-powered assault rifles are used against innocent civilians, old people, children, women. When the efforts to blame self-defined terrorists fail, mental illness is blamed, arming teachers and priests is suggested, anything but facing the fact that NRA's profit margin, not rights or justice, remains the deciding factor.

[10] Bullard, R.D., 1994, *Dumping in Dixie*, Westview Press, Boulder, CO.
[11] 424 U.S. 1 (1976).
[12] Westra (2019).

Although at Canada's borders clear signs proclaim that guns are not allowed in that country, unfortunately smuggled American guns are constantly used by disaffected youths in large cities, copying their southern neighbors. It is perhaps surprising that, just as we noted, Walmart, by no means a bastion of moral behavior, stopped selling Monsanto's Roundup, motivated by the fear of joining Bayer in its astronomical court losses over that product, and also eventually banned assault rifles from its stores. Perhaps even the profit motive may eventually suggest the change, but too many of us may need to demonstrate our status as possible victims; many even become corpses before such possible change in corporate behavior may occur. Ugo Mattei says:

> The judicial model based on checks and balances and on a division of powers has been substituted for by a model of governance concentrating all significant power in the executive. An alternative legal system or a "shadow constitution", capable of excluding judicial supervision on all political relevant issues has thus been constructed in the post-9/11 US legal system.[13]

4.1.2 Recent Corruption in Public Policy

> At ALEC (American Legislative Exchange Council) conferences, multinational corporations and state legislators meet behind closed doors to discuss, vote on, and disperse model bills. These bills increase corporate power and lower accountability while seizing the rights of ordinary citizens. The collective power of corporations such as Goldman Sachs, Exxon Mobil, and Chevron—not to mention the Koch brothers—gives them a powerful voice with business-friendly legislators who join ALEC. ALEC's Energy, Environment and Agriculture Task Force works to eliminate citizens' ability to protect themselves from pollution and other environmental threats.[14]

The 2011 Supreme Court decision essentially absolving all corporate energy persons from responsibility regarding harms arising from climate change was based on the fact that the judges claimed to lack the scientific expertise required to hand down the appropriate informed decision. The responsibility for those decisions returned to the Environmental Protection

[13] Mattei, Ugo and Nader, Laura, 2009, *Plunder*, Blackwell Publishing, p. 181.

[14] Martin, Martin, Melanie, J., "ALEC Lets Corporations Craft Environmental Policy", 29 July 2012; see http://www.earthtimes.org/green-blogs/green-opinions/alec-lets-corporations-craft-environmental-policy-29-Jul-12/

Agency (EPA) which ALEC acknowledges in its own literature (EPA's Regulatory Trainwreck, 2001), "that it aims to undermine EPA regulations" to further "the interests of the industry".[15]

Bills supported by ALEC provide an "umbrella-like protection" for industry, against all possible regional or other regulations that might propose better collaboration with climate change initiatives even at the federal level; ALEC even proposes banning all regulation of carbon dioxide. A further example is the "Model Bill" "resolution to Retain State Authority over Hydraulic Fracturing" which describes hydrofracking as follows:

Hydraulic fracturing is a proven technology with a long history of environmentally unsafe use in the completion of oil and gas wells.[16]

The proposed model bill also allows fracking companies "to avoid disclosure of chemicals". These efforts demonstrate once again a total lack of concern for human life and health, as those issues are not even factored in any of their efforts. Environmental education is also considered in their bills, some of which discourage "propagandizing" or political action and explicitly *exclude* professionals with an environmental education from their panels.[17] Is it possible that the US Supreme Court would not have been aware of all these activities contrary to the protection of human rights? It seems extremely doubtful that US judges would be unaware of the qualifications of EPA's bureaucrats, let alone their propensity to grant permits in the interests of corporations that stand behind their government's politicians.

Similarly ALEC works "to eliminate the idea of the commons-public land and resources that all people share"; its "Right to Farm Act" would even ban public objections to factory farms that pertain to pesticide use.[18] Regarding agriculture, in fact, there are no less than three separate but interconnected forms of attack that aim to ensure corporate control:

1. The ongoing flow of bills intended to diminish or eliminate corporate accountability, including a special "opportunity to correct" act, which allows companies that are discovered to be in violation of pollution regulations to forego any liability for their crimes, once

[15] ibid. p. 2.
[16] ibid.
[17] ibid.
[18] ibid.

they admit their "mistake", especially in regard to extracting industries.

2. Discouraging citizens' protests while continuing to engage in their life-threatening practices that place all sources of safe food under attack.
3. Finally, they have the option of engaging in direct acquisitions of whatever good or safe food businesses they have not been able to eliminate through the means described in 1 and 2 above.

The organic industry is the main target of the third program of attacks, and their corporations can practice what they do best: buy and sell. Philip Howard explains their strategy.[19] Most socialist theorists propose and recommend turning away from globalized corporate organizations, whether for food or other commodities. Those who prefer natural food and vegetarians tend to gravitate toward various local organic brands. Each region and state in the US have several such businesses that provide everything organic, from ice cream to beans and pasta, as well as whole frozen meals.

Howard says that "most acquisitions of organic processors occurred between December 1997, when the draft USDA standard was released, and its full implementation in October 2002[20]" as he shows how the top 30 food processors in North America have acquired organic firms; some have even "been developed specifically for Wal-Mart":

It is hard to say what remains of the organic growth/preparation promised by the USDA certification after these acquisitions take place. Organic foods are intended for the consumption of safety and health conscious individuals, but they are also—at the same time—intended for the protection of the ecology from hazardous substances. Therefore, they are part of a movement for the defense and protection of nature and life, in contrast with the corporate aims we have described, although the recent turnaround by Walmart indicates a safer, for them, if not better policy in action.

4.1.3 *The Right to Water in Wealthy Countries and Beyond*

Governments at all levels are increasingly starved for funds. Many municipal governments in the U.S. and abroad seek to obtain the funds necessary to

[19] Howard, Philip H. 2009, "Consolidation in North American Organic Food Processing Sectors 1997–2007", International Journal of Sociology, Agriculture and Food, 16(1), 13–30.

[20] Howard, Philip, Aug.4, 2012 at http://www.msu.edu/-howardp/organicindustry.html

> maintain and operate their water systems by privatizing the system: is usually promises are made of better service for lower cost. Usually the result is poorer service for higher cost.[21]

The deprivation of water does not only happen indirectly, through the promotion of inappropriate diets, but also directly as many US cities stop dealing with the public and providing water to the citizens. Privatizing water distribution demonstrates that at the root of all water problems, one finds the same disregard for the protection and promotion of public health that one finds in the production and distribution of food. The crisis is ongoing in 2016, although it initiated in April 2014, when the water contamination "was first discovered, after Flint stopped receiving its water from the Detroit Water and Sewerage Department, switching instead to water from the Flint River".

The latter system was using corroded pipes, which caused lead to leach into the water supply:

> Due to the change in water source, the percentage of Flint children with elevated blood-lead levels may have risen from about 2.5% in 2013 to as much as 5% in 2015.[22]

This was the most serious and widespread result of the public health emergency (as acknowledged by Michigan Governor Rick Snyder, on January 5, 2016), but the presence of coliform bacteria and a legionnaire's disease outbreak were also part of the health disaster, further acknowledged by President Barack Obama as a federal state emergency.

The change in water supply arose after the Flint authorities decided to save about $5 million in less than two years, by treating the water from the Flint River instead of purchasing Lake Huron water from Detroit's system, in April 2014.[23] Soon residents began to complain about the water:

[21] Dellapenna, Joseph, 2017, "Flint's Water crisis: Profitability, Cost-effectiveness, and Depriving People of water", in press.

[22] Hanna-Attisha, Mona, Lachance Jenny Sadler, Richard Casey, Champney Schrepp, Allison, 2015-12-21, "Elevated Blood Lead Levels in Children associated with the Flint Water Crisis: A Spatial Analysis of Risk and Public Health Response", *American Journal of Public Heath*, 106(2), 283–290.

[23] Winston, Samuel, October 7, 2015, "How the Flint water Crisis Emerged", *Flint Journal*, p. 2.

its color, taste and smell were not normal. Because of the presence of coliform bacteria in the Fall of 2014, a boil-water advisory was issued.

The representative of the Michigan Department of Environmental Quality (DEQ) assumed that appropriate action would be taken, but even a representative of a General Motors plant in Flint stated that the water was corroding car parts and could not be used.[24] In August 2014, THMs, a chlorine by-product used for disinfecting water, showed unsafe levels: the substance is carcinogenic and connected to other illnesses.[25]

Despite the monitoring evidence, officials indicated to the governor that "there was no imminent threat to public health", just a failure in communications.[26] It took the efforts of several local organizations, citing the presence of orange water, as well as that of numerous health effects, to accumulate and deliver more than 26,000 signatures to Mayor Dayne Walling, demanding that the city should be connected again to the Detroit water systems.[27]

It is still unclear to this day when and if the water in Flint will return to being safe for Flint citizens, despite the funds provided by Michigan legislature, the Mott Foundation and the State of Michigan itself.[28]

Dellapenna concludes his analysis of the Flint's water issues by asking what we can learn from that crisis: the first problem we must acknowledge is that the difficulties we have described are not unique to Flint, but occur repeatedly in many cities and areas of the US; second, the institutions/authorities charged with controlling the safety of drinking water and public health "have been starved of funds, just like the municipal water systems"[29]; third, a large part of US legislators and politicians from the Republican party aim to show that the Environmental Protection Agency

[24] Lin, Jeremy C.F., Rutter, Jean, Park, Hayoun, Jan.21, 2016, "Events that led to Flint's Water Crisis", *New York Times.*

[25] Fonger, Ron, June 4, 2014, "City warns of potential health risks after Flint Water tests revealed too much disinfection byproduct", *Flint Journal,* Mlive Media Book.

[26] Lin et al. (2016).

[27] Fonger, Ron, Jan.2, 2015, "Lead leaches into very corrosive Flint drinking water, viewed Aug. 2, 2016 at https://www.mlive.com/news/flint/2015/09/new_testing_shows_flint_water.html, Mlive Sept.2, 2105; "Flint returning to Detroit water amid lead concerns".

[28] Bruch, Mark, Gov. Snyder moves to come up with $12 million to switch Flint's water back to Detroit's supply, viewed Aug.2 at http://michiganradio.org/post/gov-snyder-moves-come-12million-switch-flints-water-back-detroits-supply, Michigan Radio Oct.8, 2015.

[29] Dellapenna (2017).

(EPA) and other related institutions are failing their task and should be eliminated, like many other forms of governmental regulations; and, finally, the market or cost-benefit analysis should decide how much to spend on the protection of water.

It is telling that Governor Snyder's discussion of the problems arising from lead exposure manifests a total disregard for the severity of the problem, especially as it affects young children. Bruce Lanphear has been studying childhood lead poisoning since 1904, and he states that "neurobehavioral toxicity" shows a consistent pattern of "lead-induced abnormalities":

> Antisocial behaviours such as conduct disorders, delinquency and criminal behaviors – are important components of this pattern.[30]

He adds:

> Lead exposure is a risk factor for attention-related behavioral problems and for attention-deficit/ hyperactivity disorder (ADHD), including poorer measures of attention and hyperactivity.[31]

Lead is also a reproductive toxicant. It has been associated with spontaneous abortions and low birth weight.[32]

These scientific facts are far from the simplistic assessment of the harms of lead exposure offered by then Governor Snyder: "It's only a few IQ points; it's not the end of the world".[33]

We can add yet another noxious aspect of Flint's water crisis that is similar to what we have noted in the provision and distribution of food everywhere. The nutritional and health aspects of food are seldom considered, and cost-benefit analyses or simply profit considerations take their place. In Flint, water that should eliminate thirst safely is counterproductive instead as it inflicts grave harms. Like food, which is neither examined nor controlled by impartial authorities/institutions, as it does not eliminate

[30] Lanphear, Bruce, P., 2013, "Lead", in Landrigan, P. and Etzel, Ruth, op.cit., pp. 262–264.

[31] ibid. see also Grandjean, P., 2013, *Only One Chance*, Oxford University Press, Oxford, Chapter 3.

[32] Lanphear (2013: 265).

[33] Dellapenna (2017).

Hunger but inflicts insidious harms instead. At most, its quantity is discussed, its quality is not.

For a case in another country where poverty reigns instead, Vandana Shiva explains the situation in India. Agribusiness is prepared to use any means to advance its interests, including the corruption of public officials in that country. In a short piece entitled "Make Monsanto Pay", she says:

> The Competition Commission of India (CCI), the country's antitrust regulator, has recently said that it suspects Monsanto joint venture abused its dominant position as a supplier of genetically modified (GM) cotton seeds in India and has issued an order citing prima facie violations of Section 3(4) and 4 of the Competition Act to be investigated by CCI's Director General.[34]

Shiva adds that Monsanto has treated the laws of India "as mere hurdles in its way to swindle India and its farmers", as it smuggled 100 grams of cotton seeds "containing the MON531-Bt.gene into India without the approval of the Genetic Engineering Appraisal Committee (GEAC)". Unfortunately "GMO traits, once released into the environment, cannot be contained or recalled",[35] thus allowing Monsanto-Mahyco to establish large open-field trials across nine states of India, bypassing the required approval by GEAC: the following year, in many states, the trial fields were used to plant wheat or other plants.

Of course, the advantages to Monsanto were huge: they charged farmers 80% of the cost of Bt cotton seeds (Rs 1600 per 450-gram package of Bt cotton seeds), and Rs 1250 was charged by Monsanto-Mahyco Biotech (MMB) as a "trait fee". Hence, farmers were forced to spend for the expensive seeds as well as the chemicals their genetically modified organisms (GMOs) required, thus losing gradually any hope of surviving on their lands:

> Of the 300,000 farmer suicides in India since Monsanto smuggled the Bt gene into India in 1995, 84 per cent, almost 252,000 are directly attributed to Monsanto's Bt cotton.[36]

The government of India is now suing Monsanto for "breaking Indian laws and corrupting" their regulatory system. Of course, this form of

[34] Shiva, Vandana, 2016, op.cit.
[35] ibid.
[36] ibid.

"corruption" goes hand in hand with the familiar "explosion of cancers where glyphosate is used", while, Shiva adds, the Roundup herbicide has failed across the world at controlling weeds, creating superweeds instead.[37]

Shiva addresses the grave problem of water as well, as she remarks that:

> ...modern humans have abused the earth and destroyed its capacity to receive, absorb and store water. Before station and mining have destroyed the ability of water catchments to retain water. Monoculture agriculture and forestry have sucked ecosystems dry. The growing use of fossil fuels has led to atmospheric pollution and climate change, responsible for recurring floods, cyclones and droughts.[38]

But water, she argues, is a "commons" and India passed a Water Act of India in 1974, but all the cases that appeared in the courts tended to favor the corporate polluters, rather than the people's right to clean water. She adds that, essentially, "climate injustice is water injustice"[39] and that injustice initiates with corporate control and is funded by the World Bank, which is "an instrument of corporate control over water":

> Not only has the World Bank played a major role in the creation of water scarcity and pollution, it is now transforming that scarcity into a market opportunity for water corporations. The World Bank has outstanding commitments of about 20 billion in water projects, 4.8 of which are for urban water and sanitation, 1.7 billion for rural water schemes, 5.4 billion for irrigation, 1.7 billion for hydropower and 3 billion for water related environmental projects. South Asia receives 20 percent of World Bank water loans.[40]

As well as this form of support for corporate enterprise, privatization completes the picture of water theft through corporate control. Water is also basic to agriculture, so unsustainable agriculture using GMO crops completes the devastating corporate takeover; hence climate change and starvation are indirectly connected to the dire situation of water. India's situation is not unique, but it offers an example of the intimate connection

[37] Shiva Vandana, 2016 "Make Monsanto Pay", *Commondreams*, Feb.24, viewed August 19, 2019 at http://www.commondreams/org/views/2016/02/24/make-monsanto-pay

[38] Shiva 2002, op.cit., p. 2.

[39] ibid.p. 40.

[40] Shiva, op cit, pp. 87–88.

between starvation and climate change, confirmed by the ecology we have discussed.

The issue is also clear on the other side of the world, where the burning of the Amazon is a beacon attesting to the close relation between climate change and starvation, engineered and promoted by the corporate US. Once again, like in the case of the US Constitution discussed in the previous section, established legal regimes are used, but they are perverted to suit the corporate agenda.

Business transactions are not crimes, but there are certain conditions that must be met before they may be considered legitimate. The first condition is that each part must be fully informed regarding all aspects of the pending contract. Normally, independent legal representatives are employed in order to ensure that these conditions are met. The second condition is that the negotiations should proceed under conditions of complete parity and understanding, something that is clearly impossible when one party to the agreement is an illiterate forest community, who are therefore totally unprotected in the deal. Therefore, it is futile to view these transactions as normal business transfers of property: it is far more apt to term them, as I have done elsewhere, crimes against humanity, and we will discuss further the implications regarding Indigenous peoples in Chap. 5.[41]

In this chapter we need to show clearly how these land grabs in the Amazon and other South American countries combine extreme deprivation from both the absence of availability and the production of unsafe and unhealthy food with climate change. The total absence of food affects the local populations, and the altered food production affects the populations of the world as a whole. For the former, the whole area of South America is constantly affected. Stefano Liberti describes the grave situation in the area of Mato Grosso do Sul in Brazil:

> The landscape is green but flatlthere are no trees. Only plantations that extend as far as the eye can see. Once this was a *cerrado*, a tropical environment similar to a savannah, an ecosystem with a high degree of biodiversity. Today it is mostly soya cultivation.[42]

[41] Westra, L., 2018, *Environmental Disasters and Land Grabs as Crimes Against Humanity*, Nova science Publishers, New York.

[42] Liberti, Stefano, 2013), *Land Grabbing*, Verso Publishing, London, pp. 144, ff.

The local population was Guarani Indians, and they describe those who clearly came not to conduct a business transaction, but to colonize the land. Their tractors and chainsaws were far more dangerous than bombs or guns. The Guarani had lived on their lands from time immemorial, but they had no legal title to their land, so their ownership could be brought into question. The corporate takeover was not even limited to food, as it also included investments in "etharilo", which, Liberti adds, they view as "the fuel of the future".[43]

In the Summer of 2019, as discussed in "Part 1: The Role of Lands and Forests" of Chap. 3, the burning of the Amazon was the focus of international attention. We now turn our attention to the effects of the Amazon fires on the local populations. Alexander Zaitchik addresses that issue in the "Red Zone",[44] which describes the case of the Apurina Indians of Brazil's southwestern Amazon. Several delegates of Apurina were convening by boat to discuss their emergency. In recent years the land grabbers, known as *grileiros*, or locusts, push further and further into the depth of the Amazon:

> the invasions are worse and will continue to get worse – said Francisco Umanari, a 42-year-old Apurina chief-his (Bolsonaro's) project for the Amazon is agribusiness. Unless he is stopped, he'll run over our rights and allow a giant invasion of the forest. The land grabs are not new, but it's become a question of life and death.[45]

The land grabbers want to create pastureland from the rarest, and they hack "hundred of miles through the forest with machetes and used motorcycles to transport long-blade chainsaws and drums of gasoline" (ibid., p.6). The "Red Zone" refers to agribusiness's frontier that now moves deeper and deeper into the forest: a Brazilian research center reports that in the first month of 2019 deforestation jumped more than 50 percent compared to the same period in 2018.

We will return to this topic in Chap. 4 as Zaitchik's detailed analysis of the position of the Indigenous peoples of the area will be discussed

[43] Liberti, op.cit. p. 152.

[44] Alexander Zaitchik, July 6, 2019, "Rain Forest on Fire (On the Front Lines of Bolsonaro's War on the Amazon, Brazil's Forest Communities Fight Against Climate Catastrophe") (https://theintercept.com/2019/07/06/brazil-amazon-rainforest-indigenousconservation-agribusiness-ranching/ viewed on Aug. 26, 2019.

[45] op.cit. p. 5.

together with other related material regarding people forced to abandon their homes and land in many other countries, facing the same forced eviction. In this chapter we need to return to the impact of corporate power and the results of that unchecked power, wielded with imp unity, and resulting in both climate change and starvation.

In Brazil, "5 percent of the 63 million hectares of land being cultivated in the country"[46] is lost to local people, as agribusiness surges ahead with the local extensive monocultures, as they compete with the local "sugar barons" for the most valuable asset that exists: the land. Nor is Brazil the only locus for the quest for land, as Tanzania, in Africa, is increasingly known as "the frontier for biofuels".[47] The biofuel plant is *Jatropha*, a plant that grows with minimal need for water, while its seeds remain its most valuable part.

The village chief explains how the land was secured. In 2006 a government official came to the village, asking that it should give up most of its land to a businessman, whose project in turn would build "a hospital, a school, a water pump, roads. It will also give compensation money to all the families. And it will create new jobs".[48] Only the last promise was kept, from backbreaking work in the hot sun for the equivalent of about 54 euros a month.

But the worse aspect of the "deal" was that the business, they were told, was interested in the land of eleven villages, most of which had already signed up: they were the only ones holding out. After they had signed, they found out that the same trick was used with all the other villages as well. By then Sun Biofuels had taken over 1705 hectares, but since the villagers has no legal documents to attest to their ownership, more land could be taken at will and without their approval.[49]

Nor is Tanzania the only African land that attracts land grabbers: British Sugar is carving "a whole new sugar empire in Africa", directed by the "Weston family from Canada", making them an example of "major African land grabbers".[50] These corporations are not only land grabbers; because sugar cane requires large amounts of water, they are equally "water grabbers". Another large corporation, Illovo, which started in the 1990s, has

[46] ibid.
[47] Liberti (2013: 171).
[48] Liberti (2013: 172).
[49] Liberti (2013: 173).
[50] Pearce, Fred, 2012, Beacon Press, Boston, MA, p. 238.

been buying up "farms in Malawi, Mauritius, Zambia, Tanzania and Mozambique" as well. Further Illovo purchased British Sugar, which allows it to access the European Union.[51] Thus it is not only that corporations make the food—as much as it is in their power—inadequate for its role to supply adequate nutrition and unable to truly satisfy hunger.[52] Even before cultivation and distribution, the crime of land grabbing is also committed.

It is easy to identify the deprivation of the means for survival for the whole population who lose the products of their land, hence face hunger or famine with an environmental disaster. It represents a grave, diffuse breach of human rights to life and health, as being malnourished makes people, especially the most vulnerable, prey to various deadly diseases. Water deprivation is equally grave, and it can also fit the definition of environmental disaster and a crime against humanity, given that the present accepted and permitted corporate practices and foster and support both forms of deprivation.

Hence the corporate/US engineered and supported land grabs are only the opening act—so to speak—to the one central focus, the deadly effects of climate change and starvation, of which several South American countries are well aware. On September 7, 2019, Colombia, Brazil, Bolivia, Ecuador, Guyana, Peru and Suriname signed the "Pact for the Amazon".[53] It is significant that these countries acknowledge that "What happens in the Amazon affects not just our countries but also the entire world"; these words from President Martin Vizcarra of Peru show a level of understanding that Western leaders apparently cannot match. That group's only political move was the omission of President Nicolas Maduro of Venezuela from the meeting, and even that choice was criticized by those who joined to sign "the Pact of Leticia": the main focus was protecting Mother Earth, a goal that goes beyond ideological differences.

In contrast, the question of starvation, while clearly acknowledged by the people affected in the Amazon, as well as those in the surrounding regions, is still not acknowledged by the rest of the world's leaders; even when they pay lip service to the disaster of climate change, they refuse to accept the reality of starvation, not only as a local problem but as a global

[51] ibid.

[52] UNGA/HR /34/48; Westra, L. 2017, *On Hunger*, Brown Press, FLA.

[53] https://www.telesurenglish.net/news/Seven-South-American-CountiresAgree-to-Protect-Amazon-Forests-20190907.html

issue. Our understanding is based on the fact that food that is laced with toxic materials, thus deprived of safe nourishment, cannot be accepted as a means of alleviating hunger: it is instead the root of what I have termed Hunger, and it is the root of many diseases both physical and mental, all of which not evident when the food is consumed, as they develop more or less slowly throughout the life of all who consume such food.[54]

That is a disaster that has not been acknowledged, but that has originated, at least in part, from the Amazon itself, then spread throughout the world as corporations that worked in that area are the very same who have become strengthened and enriched there and have been gaining power throughout the world, primarily in North European countries, beyond the Mediterranean countries where the pride in their own food traditions and the belief in the Mediterranean diet recommended by the WHO prevented the complete takeover of their healthy food.

It is said that, although not only the WHO has been speaking clearly about the problems engendered by agribusiness, the corporate crime that has created that additional crime against humanity has not been recognized, except in the Reports of Olivier de Schutter, which we already discussed in detail in Chap. 3. The disaster is a worldwide one, as the regions of North Africa are equally affected by corporate land grabbers.

> The oligopoly of the chemical industry has enormous power. Recent mergers have resulted in just three powerful corporations: Monsanto and Bayer, Dow and Dupont, and Syngenta and Chemchina. They control more than 65 percent of global pesticide sales. Serious conflicts of interest issues arise, as they also control almost 61 percent of commercial seed sales. The pesticide industry's efforts to influence policymakers and regulators have obstructed reforms and paralysed global pesticide restrictions globally.[55]

In order to continue to produce the huge profits of the "oligopoly of the chemical industry", the powerful corporations here named need to

[54] Tamburlini, G., 2002, "Children's Special Vulnerability to Environmental Health Hazards", in G. Tamburlini, O. von Ehrenfeld and R. Bertollini, 2002, *Children's Health and the Environment: A Review of the Evidence*, EEA Report No. 290 WHO, Geneva, Switzerland, WHO 2012, *State of the Science of Endocrine Disrupting Chemicals 2012: World Health Organization Summary for Decision Makers*, World Health Organization, Geneva; as well as the two works previously cited from Oxford, Philippe Grandjean 2013, and Philip Landrigan and Ruth Etzel, 2013, both from Oxford Publishing.

[55] A/HRC/34/18, para. 86.

ensure they control the land for their monocultures, their seeds and their genetically modified organisms. Whether the corporations grab their land they want directly or prefer to deal through the government officials of various target countries, the ultimate results persist. The bureaucracies of the countries they target ensure that laws are enacted that are favorable to those enterprises and their goals. Thus, it is hard to find domestic laws capable of restraining land grabbers, and this difficulty exists in most countries, even in those which are most affluent and secure in their geo-political position.

The US is a prime example, as its federal legal system actually appears to equate its government with one of the "oligopolies", Monsanto. The transfer of power from state-controlled agricultural expansion to "corporate food regime" has been foundational to the present situation of land grabs:

> take nineteenth century settler state agro-exporting anticipated a general offshoring of industrial agriculture, as the US-centered food regime morphed into a corporate food regime, anchored by the WTO (1995) as it institutionalized liberal trade and /investment relations. It is this trajectory that has laid the foundations for the contemporary offshore land grabs.[56]

McMichael terms the new development in land grabbing, a "new territorialization", intended to use land offshore, in order to eventually repatriate the food or other required product grown offshore.[57] This approach is "designed to curb dependence on markets" as exemplified by China's direct agreement with Argentina to secure millions of acres for their use:

> In Africa, Beijing has 1.5 million nationals building infrastructure and digging water-canals and pursuing irrigation schemes in Mali….manning seed labs in South Africa…tilling soil from Senegal to Mozambique.[58]

In addition, there is "the expanding role of agrofuels in global land grabs" as "the EU green fuel mandates [is]stimulating palm oil expansion in

[56] McMichael, Philip, 2014, "Land Grabbing as Security Mercantilism in International Relations", in Matias Margulis, Nora McKeon and Saturnino M. Borras eds., *Land Grabbing and Global Governance*, Routledge, pp. 47–64, p. 49.

[57] ibid. p. 48.

[58] Pearce 2012, *The Land Grabbers: The New Fight Over Who Owns the Earth*, pp. 202–203.

Malaysia and Indonesia, in Guatemala and Colombia".[59] Both food and energy shortages go hand in hand, and "technification" is viewed as the answer, so land grabs in faraway areas such as sub-Saharan Africa appear to offer substantial profits. This movement is even presented as a laudable effort at "feeding the world".[60]

Oxfam produced an important briefing paper on the topic "Land and Power" (The Growing Scandal Surrounding the New Wave of Investments in Land),[61] where land grabs are defined as land acquisitions which do one or more of the following:

> * Violate human rights, and particularly equal rights of women;
> * Flout the principle of free, prior, and informed consent of the affected land users, particularly indigenous peoples;
> * Ignore the impacts on social, economic, and gender relations, and on the environment;
> * Avoid transparent contracts with clear and binding commitments on employment and benefit sharing;
> * Eschew democratic planning, independent oversight, and meaningful participation.[62]

This Oxfam document states that rather than collaborating to ensure that poor people had the natural resources to which they were entitled, "too many investments have resulted in dispossession, deception, violation of human rights, and destruction of livelihoods".[63] The document considers in detail land grabs in Uganda, Indonesia, Guatemala, Honduras and South Sudan.[64] Regarding the lack of prior, informed consent, the World Bank had a special complaints mechanism, to allow those who felt a project funded by the Bank had not treated fairly a community and had not secured the community's prior informed consent.[65]

[59] McMichael (2014: 52).

[60] Peluso, N.L. and Lund, c., 2014, "New Frontiers of Land Control", The Journal of Peasant Studies, 38(4), pp.667–681, pp.673–4.

[61] 171 Oxfam Briefing Paper – Summary – September 22, 2011, henceforth Oxfam, 2011.

[62] ILC (2011) "Tirana Declaration" Securing Land access for in times of intensified natural resources competition", International Land Coalition, viewed July 12 at http://www.Lnadcoalition.org/about-us/aom/2011/tirana-declaration

[63] Oxfam (2011).

[64] ibid.

[65] Herbertson, Kirk, Thompson, Rim and Goodland, Robert, 2010, *A Roadmap for Integrating Human Rights into the World Bank Group*, Washington, DC, World Resources Institute.

Margulis and Porter acknowledge this grave difficulty as the emphasis on simple "transparency" tends to advocate "reliance on a process of information dissemination which obscures the deficiencies of that process itself",[66] while the same process also obscures the difference between disparate groups, thus hiding the complex relations that produce exploitative transaction that define land grabbing.[67]

The authors argue that the recent circumstances within which land grabs take place indicate a complete change from the earlier geopolitical situation, characterized by the clear North/South and West/East conflicts and dominated by the power of US hegemony which followed after WWII, the era of colonialism of the European sovereign states, such as Spain, the Netherlands, France and Great Britain. During that period US power, the authors believe, included the power of multinational corporations, like "Monsanto and Cargill".[68] But in recent times, the geopolitical map is substantially changed because of the input from many other actors as the sources of capital for "food-feed-fuel" now include private investors, the powerful agri-food business, deregulated through global trade, and the developing of land as a new asset. Land grabs now often proceed through novel formal practices.

While these changes are undeniable, it seems that rather than replacing the obvious relations between the "US-led capitalist West and state-centric regions elsewhere", the newer complexity is superimposed, added to the existing forms of imperialism. The reason for this suggestion is that the increasingly evident trend is the actual identification of the US itself with the agribusiness multinational corporations it supports. In turn the latter control in various ways the legal regimes originally intended to control their activities in protection of the public interest, either through economic means or through direct intrust on the regulatory process.

References

421 US 1, 96S.ct. 612, 46L
424 U.S. 1 (1976)
A/HRC/34/18, para. 86

[66] Margulis and Porter (2014: 73).
[67] ibid.
[68] Margulis and Porter (2014: 70).

Alexander Zaitchik, July 6, 2019, "Rain Forest on Fire (On the Front Lines of Bolsonaro's War on the Amazon, Brazil's Forest Communities Fight Against Climate Catastrophe)", https://theintercept.com/2019/07/06/brazil-amazon-rainforest-indigenousconservation-agribusiness-ranching/ viewed on Aug.26, 2019

Barlow, Maude and Clarke, Tony, 2003, *Blue Gold* (The Battle Against Corporate Theft of the World's Water), "Losing the Great Lakes" pp 35–6, McLelland and Stewart, Toronto, Ont.

Bruch, Mark, Gov. Snyder moves to come up with $12 million to switch Flint's water back to Detroit's supply, viewed Aug.2 at http://michiganradio.org/post/gov-snyder-moves-come-12million-switch-flints-water-back-detroits-supply, Michigan Radio Oct.8, 2015

Bullard, R.D., 1994, *Dumping in Dixie*, Westview Press, Boulder, CO

Dellapenna, Joseph, 2017, "Flint's Water crisis: Profitability, Cost-effectiveness, and Depriving People of water", in press

Dockrill, Peter, *Science Alert*, June 26, 2019, viewed Sept. 22, 2019 at www.businesainsider.com/climate-apartheid-united-nations-report2019-6?IR=T

Fonger, Ron, Jan.2, 2015, "Lead leaches into very corrosive Flint drinking water, viewed Aug. 2, 2016 at www.mlive.com/news/flintindex.ssf/2015/09/new_testing_snows_flint_water.html, Mlive Sept.2, 2105; "Flint returning to Detroit water amid lead concerns"

Fonger, Ron, June 4, 2014, "City warns of potential health risks after Flint Water tests revealed too much disinfection byproduct", *Flint Journal*, Mlive Media Book

Grandjean, P., 2013, *Only One Chance*, Oxford University Press, Oxford, Chapter 3

Hanna-Attisha, Mona, Lachance Jenny Sadler, Richard Casey, Champney Schrepp, Allison, 2015-12-21, "Elevated Blood Lead Levels in Children associated with the Flint Water Crisis: A Spatial Analysis of Risk and Public Health Response", *American Journal of Public Heath*, 106(2), 283–290

Herbertson, Kirk, Thompson, Rim and Goodland, Robert, 2010, *A Roadmap for Integrating Human Rights into the World Bank Group*, Washington, DC, World Resources Institute

Howard, Philip H. 2009, "Consolidation in North American Organic Food Processing Sectors 1997–2007", International Journal of Sociology, Agriculture and Food, 16(1), 13–30

Howard, Philip, Aug.4, 2012 at http://www.msu.edu/-howardp/organicindustry.html

ILC (2011) "Tirana Declaration" Securing Land access for in times of intensified natural resources competition", International Land Coalition, viewed July 12 at http://www.Lnadcoalition.org/about-us/aom/2011/tirana-declaration

Landrigan, Philip and Ruth Etzel, 2013, *Textbook Children's Environmental Health*, Oxford University Press, New York

Lanphear, Bruce P., 2013, "Lead", in Landrigan, P. and Etzel, Ruth, op.cit., pp.262–264
Leopold, Aldo. 1933, "The Conservation Ethic", Journal of Forestry, 31, 6
Liberti, Stefano, 2013, *Land Grabbing: Journeys in New Colonialism*, Verso Publishing, London, pp.144, ff.
Lin, Jeremy C.F., Rutter, Jean, Park, Hayoun, Jan.21, 2016, "Events that led to Flint's Water Crisis", *New York Times*
Margulis and Porter, 2014:73
Martin, Melanie J., "ALEC Lets Corporations Craft Environmental Policy", 29 July 2012
Mattei, Ugo and Nader, Laura, 2009, *Plunder*, Blackwell Publishing, p. 181
McMichael, Philip, 2014, "Land Grabbing as Security Mercantilism in International Relations", in Matias Margulis, Nora McKeon and Saturnino M. Borras eds., *Land Grabbing and Global Governance*, Routledge, pp.47–64, p.49
Oxfam Annual Report, 2011
Pearce, 2012, *The Land Grabbers: The New Fight Over Who Owns the Earth*, pp.202–203
Peluso, N.L. and Lund, c., 2014, "New Frontiers of Land Control", The Journal of Peasant Studies, 38(4), pp.667–681, pp.673–4
Shiva, Vandana, 2002 *Water Wars* (Privatization, Pollution and Profit), Between the Lines, Toronto, Ontario
Shiva, Vandana, 2016 "Make Monsanto Pay", *Commondreams*, Feb.24, viewed August 19, 2019 at http://www.commondreams/org/views/2016/02/24/make-monsanto-pay
Tamburlini, G., 2002, "Children's Special Vulnerability to Environmental Health Hazards", in G. Tamburlini, O. von Ehrenfeld and R. Bertollini, 2002, *Children's Health and the Environment: A Review of the Evidence*, EEA Report No. 290 WHO, Geneva, Switzerland
UNGA/HR /34/48
United Nations Office of the High Commissioner on Human Rights, Statement by Prof. Philip Alston, UN Special Rapporteur on Extreme Poverty and Human Rights, Malaysia, 13–23 August 2019
Venton, Peter, 2020, The Political Economy of Managing Without Growth, in press
Westra, Laura, 2017, *On Hunger*, Brown Press, FLA
Westra, Laura, 2018, *Environmental Disasters and Land Grabs as Crimes Against Humanity*, Nova Science Publishers, New York
Westra, Laura, 2019, *From Principles to Norms: The Development of International Law*, Nova Science Publishers, New York
WHO, 2012, *State of the Science of Endocrine Disrupting Chemicals 2012: World Health Organization Summary for Decision Makers*, World Health Organization, Geneva
Winston, Samuel, October 7, 2015, "How the Flint water Crisis Emerged", *Flint Journal*, p.2

CHAPTER 5

Migrants and Refugees from South America to North Africa, Starvation and Climatic Conditions Not Acknowledged

Abstract Westra turns to the practical results that follow both scientific research advances and the powerful corporate opposition to any law that might restrict their operations, regardless of the human rights disasters that will follow. Most affected are the Indigenous peoples, and Canada has specific regulations protecting their First Nations (Fns) and so do New Zealand, Australia, and several South American nations. Yet in all instances corporate impunity wins, as those victimized by ineffectual laws attempt their desperate escapes from North Africa and the Middle East in unsafe boats, attempting to reach the closest nation in the Mediterranean to find safety. Legal regimes exist both for the protection of those arriving on land and on the high seas. But the racist attitudes they find in the countries of arrival, especially Italy, mean that populist/Fascist governments are intent on scoring points for future elections rather than respecting the law. Their position is to blame the victims for their plight, calling them complicit with those who explicit them.

Keywords Racism • Indigenous peoples • Ineffectual legal regimes on land and sea • Populism and Fascism in Italy and other Mediterranean countries

L. Westra, *Climate Change and Starvation*, Palgrave Insights into Apocalypse Economics,
https://doi.org/10.1007/978-3-030-42124-3_5

5.1 Introduction to the Issue of Necessary Escapes

> Mr. Alston said that his report on climate change focused on the relationship between climate change, human rights, poverty and inequality, and reflected an overwhelming urgency of climate change in relation to the mandate. Climate change posed a grave and immediate challenge, the Special Rapporteur warned....[1]

Further, the Special Rapporteur added that the results of climate change he observed would result in climate apartheid.[2] The report traces the ongoing situation, where the rich can adapt to climate change, the poor cannot. In addition, we will argue, it is the wealthy countries that are responsible for climate change and for the results the UN is reporting, either directly or indirectly, through the corporate activities they permit or even encourage.

However, it is not only the poor, in general, who are thus victimized, but, most often, poverty is coupled with Indigenous status which in turn involves a number of legal regimes intended to protect their rights singly and collectively. Some countries, such as Canada, have specific regulatory regimes intended to deal with their First Nations (FNs), as does New Zealand, Australia and several South American countries. The question is whether any of these existing laws are actually effective and—if they are not as seems to be the case—what are the impediments to their implementation, and where does the responsibility lie for the inefficiency of these well-meaning legal instruments.

Indigenous people are not the only victims of unrecognized forcible evictions from their homes and lands. Too many, both Indigenous peoples and other poor, end up in the closest countries to the most affected regions, that is, the Middle East and North Africa. They eventually aim for Italy, Greece, Malta and Turkey, the closest nations. Their position is unsustainable at most; they find the money to pay for a dangerous

[1] Human Rights Council Holds an interactive dialogue on Extreme Poverty and on internally displaced persons", 28 June, 2019, viewed October 11, 2019 at https://www.ohchr.org/FR/l-tRBodies/l-tRC/Pages/NewsDetail.aspx?NewsID=24765&LanglD"'F

[2] Dockrill, Peter, *Science Alert*, June 26, 2019 "We're headed for "climate apartheid' in which the poor will suffer, while the rich save themselves, warns a chilling UN Report", viewed Sept 22, 19 at https://www.businessinsider.com/climate-apartheid-united-nations-report-2019-6?IR=T

transport across the Mediterranean, but they are treated like criminals when they arrive, despite the existing laws for their protection, both on land and on the sea. At worst, they are returned to their war-torn country or the land affected by climate change from which they originated.

Once again, the presence of detailed and time-honored legal regimes offers neither protection nor security. The states where they arrive are increasingly regressing to extreme nationalism with unfortunate Fascist overtones. Italy presents a clear example of this political trend. When he was at the helm, Matteo Salvini, who was a minister at the time, now opposition representative, first signed the Dublin Accord; later, although most politicians in that country acknowledge the "Europe starts at Lampedusa" with reference to the most southern port, the target of most refugees' conveyances, he changed his attitude hoping to gain future votes from worried citizens. He started to describe refugees as criminals, at least until proven to be "innocent", and a threat to Italy's safety. Their "criminality" was based on their cooperation with those who transported them, for a price.

Italy is not the only country who attempts to justify its cruel policies regarding refugees, by blaming the victims, and we will argue that it is the current weakness of each of those states which spurs them to justify their own existence by exerting whatever limited power they still demonstrate. Their limited ability to control the situation within their borders, coupled with the weakness of international law, renders the appeal to extreme nationalism an obvious choice.

In contrast, regulatory regimes regarding refugees and migrants both on the sea and on land exist, and they are extensive and comprehensive. Their requirement cannot be answered by a generalized assumption of guilt. Nor can the presence of the original causality for their flight be ignored. Both climate change and starvation are never acknowledged in that context, and we will review the reasons for those omissions and the possible answers that may be proposed to a situation that is only acknowledged now by extremely visible protests.

These protesting millions can be seen on most television stations. But the demands of the protesters are not reflected in the choices of policymakers. In fact, the few European leaders that dare to admit the compelling reality of the protesters' cause tend to lose even the limited power they still possess to extreme right candidates. The "powerless" protesters, primarily the children, seem to gain momentum as they even dare, as we shall see, to sue their own governments for their climate inaction.

Most of those who are under attack are Indigenous peoples attempting to live on their lands, before climate disasters and corporate land grabbing destroyed their life. The legal regimes protecting Indigenous peoples are many and well-entrenched in international law, as well as in some countries' constitutions. Yet after we examine those documents, we will see that the situation of Indigenous peoples on the ground is actually not protected, as they are the bulk of the desperate refugees and the corporate criminals that are essentially the cause of their situation proceed with impunity in all continents.

Recently the fall of the Berlin Wall was celebrated with great fanfare, entailing, it was said, the triumph of our common humanity and the rights to be united. It is unfortunate that, despite the high-sounding words, walls are being built everywhere from the US to Israel to many so-called camps where refugees are imprisoned. Of course both those who construct the wall and those who manufacture the arms used to defend the walls against those attempting to survive are part of a very rich corporate number of those who pursue profit at all costs, even if it means arming one preferred country against another with no regard for law let alone justice.

5.2 Extreme Poverty and the Denial of Human Rights

The UN Rapporteur, Mr. Philip Alston, visited the UK and Northern Ireland, the Democratic Republic of Lao and the Philippines. He noted that on June 27, 2019, the city of Geneva "had set several records for maximum High temperatures", which encouraged him to add there was

> an intimate relationship between climate change, human rights, poverty and inequality, and reflected the overwhelming urgency in relation to the mandate. Climate change posed a grave and overwhelming challenge….[3]

He added that the responsibility for the present situation lies with the wealthy countries, something that no present national or international legal instrument is prepared to acknowledge. They are responsible for the present and ongoing "climate apartheid" but also, specifically, for the

[3] HCDH, "Human Rights Council holds interactive dialogue on extreme poverty and on internally displaced persons", https://www.ohchr.org/FR/HRBodies/HRC?Pages/NewsDetail,aspx?NewsID=24765&LandID=F, 28 June, 2019, p. 5.

attacks on human rights of internally displaced persons singly but also of whole communities.

Responsibility, however, is not accountability, so even those who may recognize the issues involved cannot enforce corresponding accountability, as it reflects on internally displaced persons. Cecilia Jimenez, Special Rapporteur on the Human Rights of Internally Displaced Persons, notes "the lack of political will of national authorities", despite the fact that 2018 was the twentieth anniversary of the Guiding Principles on Internal Displacement.[4] In 1993, at the end of then Representative of the Secretary General on Internally Displaced Person, Francis M. Deng acknowledged that there were many "gray areas" in the provisions on internally displaced persons in existence.

If these principles had been acknowledged, then the present misnaming and mistreatment of those who are forcibly evicted by circumstances beyond their control from their land of origin would not take place. Climate change plays a pivotal role in their existence. We need to keep in mind that the poor, the internally displaced, are not all, or even for the most part, in the cities, let alone working for a wage. In fact, the city-dwelling poor, although indeed disproportionately suffering from climate change effects, have not been suffering for land grabs.

Their home bases may still be theirs, but they have been rendered inhabitable because they have been the target of disproportionate pollution, through "environmental racist" practices in North America.[5] Yet, as Richard Westra suggests:

> Marx himself established his environmental credentials in his theorizing on the basic contradiction of capitalism. Use Value life and the metabolic interchange between human beings and nature that reproduces it, presupposes the ecological sanctity of nature, the biosphere and the geosphere.[6]

There are many beyond the urban poor who are severely affected by both climate change and starvation: the Indigenous peoples of the world who, as we shall see, are, at least on paper, protected much better than the

[4] E/CN.4/1998/53/Add,2.

[5] Bullard, Robert, 1994, *Dumping in Dixie*, Westview Press, Boulder, CO.

[6] Westra, Richard, August 19, 2019, "Green Designs: for 21st Century Socialism", viewed Aug 19/19 at https://socialistproject.ca/2019/08/ green-designs-for-21st-century-socialism/p.3

urban poor by law, although that is primarily international law that speaks to their rights and their protection.

5.2.1 *Indigenous Peoples and Minorities in International Law*

The first thing to note is that Indigenous peoples are often discriminated against on racial ground, even before we can consider the specific legal instruments that are present in international and constitutional law for their protection. Hence, we will start by considering the general racial aspect of their plight.

> The Committee is conscious of the fact that in many regions of the world indigenous peoples have been, and are still being discriminated against, deprived of their human rights and fundamental freedoms and in particular that they have lost their lands and resources to colonists, commercial companies and State entrepreneurs. Consequently, the preservation of their culture and their historical identity has been and still is jeopardized.[7]

Some of the most basic legal principles existing in defense of Indigenous people are to be found in the work of the *Committee on the Elimination of Racial Discrimination* (CERD). But that work and indeed that document is fairly recent, whereas the principles upon which it is based can be traced far back both in morality, in the sense of respect for the dignity of human beings,[8] and in the law.

An exemplary case of this moral principle, together with the aspiration to see it enshrined in international law, is the *Le Louis* case.[9]

On March 11, 1816, the "Queen Charlotte" cutter captured a French ship, *Le Louis*, that had sailed from Martinique, destined for Africa and back. *Le Louis* was seized and taken to Sierra Leone, where the Vice-Admiralty Court heard the pleadings of the case, to the effect that *Le Louis*

[7] *Committee of the Elimination of Racial Discrimination* (CERD), general recommendations (XXITTI) concerning indigenous peoples; Adopted by the U.N. Committee on the Elimination of Racial Discrimination at its 1235th meeting, on August 18, 1997. U.N. Doc. CERD/C/51/ misc.13/Rev.4 (1997).

[8] Beyleveld, Deryk and Brownsword, Roger, 2001, *Human Dignity in Bioethics and Biolaw*, Oxford University Press, Oxford.

[9] *Le Louis*, 2 Dods. Rep. 210; it is "the sentence of a Vice-Admiralty Court, condemning a French ship for being employed in the slave trade, and for forcibly resisting the search of the King's cruisers, reversed – No British Act of Parliament, a commission funded upon it, if inconsistent with the law of nations can affect the rights or interest of foreigners".

was clearly fitted and intended for the slave trade from Africa; that the slave trade was contrary to both the Treaty of November 20, 1815, between England and France (abolishing the slave trade) and the laws of both countries, as well as the "law of nations"; and that the crew of the "Queen Charlotte" was duly empowered to search the French vessel. The court also relied on an earlier case, "The Amedie"[10]:

> … in which it is laid down generally by the Superior Court that the slave trade is prima facie illegal, and that the burden of proof is on the claimants to shew (sic) that the laws of their own country permit such a traffic.[11]

However, Sir Walter Scott in his judgment, despite his horror at the "trade" involved, took quite a different position, as he argued:

1. that the search and seizure were both illegal according to the law of nations, as the slave trade is not considered a crime internationally, as piracy is, which alone grants leave to anyone to seize and search vessels on the high seas in peace time;
2. that because of that illegality, the fruits of the search should not have been admitted in evidence, and the force used in the search and seizure that led to the death of five persons and many injuries could not have been legal in itself;
3. that even if the *British Slave Trade* Act, 51 G3 slave trading were indeed illegal, *Le Louis* was a French vessel with French documents and a French Master, thus that it remained unclear at that time, whether the condemnation of the slave trade (initially made by Bonaparte, but after the latter's exile to Elba, also adopted by the King of France was indeed illegal). The note of the British Minister to Prince Talleyrand (the French King's minister) elicited only the information that "His Most Christian Majesty" had issued "directions" to the effect that "the traffic should cease for the present time everywhere and forever" but that a formally promulgated law appeared to be missing at that time[12] and was apparently only formalized in 1817.

[10] 1 Dods. 84.
[11] *Le Louis*, 2 Dods. 211–212.
[12] ibid., 237–259.

This is a highly instructive case, because slavery, the ultimate form of racial discrimination, was viewed with unquestionable distaste and even disgust, by the judge and the court, but, in the final analysis, it was not possible to condemn the vessel's activities, and free the slaves, while condemning the ship's Master as well as his country, as there was no law expressing forbidding that activity at the time.

Sir Walter Scott says:

> I must remember that in discussing his question, I must consider it, not according to any private moral apprehension of my own (if I entertained them ever so sincerely) but as *the law* considers it.[13]

He added that to simply consider the state of laws at the time (from one perspective, without the presence of a "United Nations" or the possibility of enforcing *erga omnes* obligations) had he attempted to condemn and punish the practice, through violent acts, would be

> ... to force the way to the liberation of Africa by trampling on the independence of other states in Europe; in short, to procure an eminent good by means that are unlawful; is as little consonant to private morality as to public justice.[14]

What is obvious in this case is that the common morality of the day, at least in England, viewed slavery as akin to piracy, a practice that rendered those who followed it *hostes humanis generis* (enemies of humankind), thus open to seizure and attacks on the open seas even in times of peace. Yet the lack of a specific law forced a judge who strongly believed in the wrongness of the slave trade to even argue that it was hard to term it a "crime", as it had historically been practiced from Ancient times to date and was even permitted in British colonies.

At any rate, one can judge the vast gulf that separates almost universally held moral principles from the legal codification of those principles into international law. The slave trade has been abolished for a long time, but racism and ethnically motivated attacks are alive and well in all regions of the world. Hence the thrust of this chapter is to see the physical reality of the harms against indigenous communities acknowledged in law. It is

[13] ibid., 247.
[14] ibid. 257.

incredible that we are still, in some sense, placed in the position of Sir Walter Scott, viewing a crime he abhorred being committed but being unable to enforce its punishment in the law of the time, of an act that his conscience told him unequivocally was against both morality and justice.

However, there are many specific legal instruments that are available to protect their rights, as Indigenous peoples are the most affected among the poor and the most defenseless despite existing instruments.

> International law is not only rules. It is a normative system harnessed to the achievement of common values, values that speak to all of us.[15]

The first thing to note is that there is not one absolute definition of "Indigenous peoples" in international law, although they are increasingly emerging both as players and participants in the UN instruments as well as other documents.[16] The main international law instrument that attempts to define Indigenous peoples and their rights is the ILO,[17] which treats the following groups as "Indigenous":

- peoples whose social, cultural and economic conditions distinguish them from other sections of the national community.
- peoples whose status is regulated wholly or partially by their own customs and traditions.
- peoples who descend from populations that inhabited a country at the time of conquest or colonization.
- self-identification of a group as indigenous or tribal is regarded as a fundamental criterion.[18]

[15] Higgins, Rosalyn, 1994, *Problems and Process: International Law and How we Use It*, Clarendon Press, Oxford, UK, p. 11.

[16] Metcalf, Cherie, 2004, "Indigenous Rights and the Environment: Evolving International Law", in Ottawa Law Review 35,101.

[17] no.169 (*ILO Convention on Indigenous and Tribal Peoples*, No. 169 of 1989, International Labour Conference, (entered into force Sep.5, 1991).

[18] Metcalf (2000), ibid., p.103; see also a UN Study that has the following definition "Indigenous communities, peoples and nations are those which, having a historical continuity with pre-invasion and pre-colonial societies that developed on their territories, consider themselves distinct from other sectors of the societies now prevailing in those territories, or parts of them. They form at present non-dominant sectors of society and are determined to preserve, develop and transmit to future generations their ancestral territories, and their ethnic identity as the basis of their continued existence as peoples, in accordance with their own cultural patterns, social institutions and legal systems". U.N. Of Discrimination and

Also, there are legal instruments that are relevant to both Indigenous peoples and their environment, but only two of these are legally binding: *The International Labor Organization (No.169) Concerning Indigenous and Tribal Peoples in Independent Countries* and the *Convention on Biological Diversity.*[19] James Anaya[20] traces the rights of Indigenous peoples to the earliest times of international law, as they emerged as a topic of discussion after Christopher Columbus's "discovery".[21]

From the writings of Roman Catholic missionaries such as Francisco de Vitoria, the maltreatment of the "other Indians" is clearly documented, as was the natural law basis for the severe critiques of those ongoing practices.[22] Despite later criticisms of natural law-based argument for human rights,[23] the natural law approach can best provide the basis for an all-embracing system of human rights protection, a system that later Grotius attempted to separate somewhat from that doctrinal origin:

> Grotius moved toward a secular characterization of the law of nature, defining it as a "dictate of right reason" in conformity with the social nature of human beings.[24]

Nevertheless, it is only because natural law claims a supranational source for its moral perspective, so that it is not simply "humanist",[25] that natural law could and did claim to be able to judge positivist laws. For natural law, a law that violated the moral code was not truly law at all.[26]

Protection of Minorities, *Study of the Problem of Discrimination against Indigenous Populations*, U.N Doc. E/CN.4/Sub.2/1986/7Add.4, para.379.

[19] 5 June 1992, 1760 U.N.T.S. 79, Can.T.5.1993 No.24,31 I.L.M. 818, henceforth *Biodiversity Convention*.

[20] Anaya, S. James, 2004, *Indigenous Peoples in International Law* sec. ed. Oxford University Press, NY.

[21] Anaya, ibid.

[22] ibid.

[23] See, for instance, Upendra Baxi, 1998, "Voices of Suffering and the Future of Human Rights", 8 Transnational Law and Contemporary Problems, 125.

[24] Anaya, ibid., p A7: see also Grotius, *The Law of War and Peace* 40, Classics of International Law ed. 1925, Francis W. Kelsey trans. of 1646 ed.

[25] *pace* Anaya.

[26] Anaya, ibid.,; see also Thomas Aquinas, who termed any law that did not promote and protect the interests of *all* citizens, a form of violence by a sovereign or law-maker; as such it was the citizen obligation not only *not* to obey it but also to actively oppose it; see Westra Laura, 2004, *Ecoviolence and the Law*, Transnational Pub., Ardsley NY, p.38; also Martin

But in the century after Grotius, Emmerich de Vattel wrote his *The Law of Nations*,[27] where he argued that natural law should simply be applied to nation/state as to individuals.[28] Perhaps this approach was foundational not only to the correct positivist law preference, but it may also be viewed as the origin of the later application of individuals' rights and norms to corporate legal entities. Both moves proved to be highly damaging to the individual rights of national person. In modern times, the prevalence of ecofootprint harms renders these rights particularly vulnerable when the individuals at risk are part of an indigenous population.[29]

In sum, the step from the natural law protection of all individual basic rights to the positivist move to subsuming individuals under the category of nation/states papers over a vast area of differences that exist, for instance, between minority groups and others, rich and poor, colonizers and colonized. The lack of recognition of these fundamental differences is highly damaging to Indigenous peoples, as they are slowly attempting to regain, singly and collectively, the rights they might have retained historically under a different conceptual understanding of the law.

In order to apply international law regimes regarding Indigenous peoples to First Nations and other Aboriginal groups, the first steps are (a) to see whether they fit under existing accepted definitions of Indigenous peoples and (b) discover what binding or suggestive legal instruments may exist that are relevant to the protection of their interests and (c) what soft law instruments may be used to advantage. In addition to International Law Covenants, it would be good to also include the *Earth Charter* with its strong component of support for all, habitats and natural entities through ecological integrity, and with its explicit support for respect for Indigenous peoples.[30]

International law, as we saw, provides protection for Indigenous peoples, contradiction of capitalism, at least on paper, yet we see little or no results. In contrast some states handle the issue in a better manner, at least

Luther King, Jr., 1990, "Letter From a Birmingham Jail" in R. Holmes ed., *Nonviolence in Theory and Practice*, Wadsworth Pub, pp. 60–71.

[27] De Vattel, Emmerich, *The Law of Nations*, Classics of International Law Series, 1916, Charles G. Fenwick trans. of 1758 ed.

[28] Anaya (2004: 20).

[29] Westra, Laura. 2006, *Future Generations Rights and the Rights of the First Generation*, Earthscan Publications, UK, Ch.7.

[30] *The Earth Charter*, The Earth Charter Initiative, International Secretariat P.O. Box 319–6100 San Jose, Costa Rica.

to the extent that they acknowledge it as a problem. Canada, for instance, has enshrined the rights of its First Nations in its Constitution.[31] In addition, Canadian scholars have been researching the Canadian Arctic.[32] C. Kendall puts it well:

> Pack ice to the white man seems like a barrier, something to fear. But to the Inuit it's their highway. It's their communication system, their freedom, their independence. Without it there is no Inuit culture.[33]

Similar concerns and corresponding protection can be found in the Australian Court.[34] These are all specific cases that indicate increasing concern for Indigenous peoples. Yet the situation appears to worsen instead, and we can seek the reason for that in the complicity of state and corporate control.

5.3 The Reality of Corporate Control

> The enterprise operates as a quasi-state whose social obligations to those under its control are accepted in both moral philosophy and international law doctrine.[35]

Unfortunately, corporations are not visibly influenced by either moral philosophy or international law: these exist as we saw in the previous sections but appear inefficient thus far. After the human rights violations of WWII, and starting with the principles of Nuremberg, the new international law instruments under the aegis of the UN prescribed individual criminal accountability for violations of the laws of war, as well as

[31] Constitution Act (1982), Schedule B of the Canada Act 1982 (UK), a part of the Constitution of Canada.

[32] Ford et al., 2006, On the Arctic Climate Impact assessment (ACIA), Ford, J, et al. "Vulnerability to Climate Change in the Arctic", *Global Environmental Changes*, Vol.16, pp. 145–160.

[33] Kendall, C., 2006, "Life at the edge of a warming world", *The Ecologist*, Vol.36, No,5, July/August, p. 26.

[34] Dagi, Shaokles, Amben, Maur and others v. Broken Hill Property Company Ltd., and the Ok Tedi Mining Ltd. (No2), (1997) Victoria Reports (VR) 428) *Mabo v. Queensland*, 83 ALR High Court of Australia, 1992, 175 CL). The same hold true for New Zealand (Maori Council v. Attorney General (1987) IN2LR (1641).

[35] Ratner, Steven R., 2001, "Corporations and Human Rights, A Theory or Legal Responsibility", *Yale Law Journal*, vol. 111, pp.443–545, p. 509.

prescribing a "corpus of law prescribing limits upon a government's conduct towards its own citizens".[36] Nor is it only Ratner and other scholars who denounce the criminality of present corporate behavior.

Pope Francis, during a speech at the Vatican, before the 20th World Congress of the International Association of Penal Law, coupled the presence/resurgence of Fascism/Nazism with "the culture of waste and hate" which go hand in hand, as politicians increasingly "spew derogatory and racism attacks" against "homosexuals, gypsies, Jewish people and others". He proposes that the Catholic Church should add a new sin, "the sin against ecology", a duty that is imposed upon us "by the urgency of climate change fostered by crimes against the environment".[37] Even a cursory glance at some of the corporate crimes, even in a country constitutionally committed to the protection of its FNs, as well as generally in favor of social protection of the poor, like Canada, will easily demonstrate the presence and impunity of corporate crime.

5.3.1 *Two Cases of Corporate Criminality in Canada*

> The right to health is a legal instrument—a crucial and collective tool for the health sector to provide the best care for patients and to hold national governments, and the international community to account.[38]

Yet, despite the existence of that "right", at least in theory, gross attacks on the life and health of human beings persist even in modern, Western, democratic countries, and Canada is a prime example of that problem. Before proceeding with examples, it is important to note that this passage unwittingly perhaps hides the main cause of ill health: not the lack of healthcare after one becomes ill, but the presence of toxic exposures that make one ill: these are for the most part perfectly legal and ongoing in developed countries, and even more unrestrained in developing ones, under the heading of "development".

[36] Ratner et al., 2009, *Accountability for Human Rights Atrocities in International Law: Beyond the Nuremberg Legacy*, 3rd ed. Oxford University Press, Oxford, UK, p. 4.

[37] Queally, Jon, Nov 16, 2019, "While warning of Nazi-like Fascism and Corporate Crimes, Pope Francis Proposes adding 'Ecological Sin' to Church Teachings", viewed Nov 19 at https://www.commondreams.org/newa/2019/11/16/while-warning-nazi-fascism-and-corporate-crimea-pope-francis-proposes-adding

[38] Editorial: "The Right to Health: From Rhetoric to Reality", *The Lancet*, online Dec.10, 2008.

Toxic Trespass was the name of a 2005 short film, the doctoral thesis in social studies, of Dorothy Goldin-Rosenberg of Toronto.[39] Technically, this is not a case, as it never appeared before a court of law. The problem is the presence of numerous cases of asthma, childhood cancers and other diseases, often present in several children in the same family.

This phenomenon was noted by grassroot groups in the city of Windsor, Ontario, Canada, and a similar picture emerged from Sarnia, Ontario, in the heart of the "chemical valley" between Ontario and Detroit Michigan, US. For Windsor, the city lies at the end of the Ambassador Bridge, which connects it with Detroit: an exponential increase in the numbers of truck traffic to and from Detroit entails an ongoing and ever-increasing exposure to all the pollutants such traffic produces.

Various Windsor groups and citizens' associations joined together to research and compare notes, involving various Toronto hospitals and health facilities in the US and even testing children in the area for the presence of toxic substances in their blood. Some of the background of the history of carcinogenic exposures can be found in Debra Davis's landmark work.[40] A segment of *Toxic Trespass* has Dr. Davis discussing the issue in person.

A similar situation was and is present in Sarnia. In the same area, the Aamjiwnaang First Nation suffers even worse attacks to their health and normal functioning, as the effluents and particulates from the operation of multiple chemical/industrial compounds operate in the area. The effects of both must be controlled the high-intensity traffic in Windsor and several industrial operations, each one of which *may*, most of the time, operate at the limit prescribed by law but whose combined and cumulative effects have never been tested until Health Canada eventually did so.

Health Canada finally acknowledged the direct relation between the body burden of pollutants in the area and a number of grave issues, beyond childhood asthma and cancers. In the area inhabited by that FN, many couples had children with cognitive and other abnormalities, stillbirths and spontaneous abortions were common, and—most ominous of all—the birthrate of male children has been in steady decline, in comparison

[39] See Scott, Dayna Nadine, 2008, "Confronting Chronic Pollution: A Socio-Legal Analysis of Risk and Precaution" 46 Osgoode Hall Law Journal, 29 for a legal discussion of the issue.

[40] Davis, Debra, 2007, *Secret History of the War on Cancer*, Basic Books, New York.

with the birth rate of females. That is the typical result of exposure to endocrine disruptors and pesticides.

In November 2006, Philippe Grandjean and Philip Landrigan published their landmark research, where they listed hundreds of commonly used untested chemicals, produced by various corporations after WWII.[41]

Children are the first and most obvious "canaries in the mine", because of their smaller size, which magnifies their exposure, and because of their specific biology and forms of development. But various chemical exposures have now been found to promote not only the main diseases mentioned thus far but also both Parkinson's and Alzheimer's, diseases of later onset.

Nor are all pesticides and other such chemicals the only culprits in diseases of later life: aluminum, a metal with which we are constantly in contact in daily life, has been found at the core of each of the plaques that indicate the increasing loss of nervous tissue (discovered in post-mortem examination), typical of the brain of those suffering from Alzheimer's disease.[42]

5.3.2 Pulp and Paper Mills and Mercury Poisoning: The Canadian Case of Grassy Narrows and White Dog

> ...the settlement and the events leading up to it provide a striking example of the fragility of Canadians' environmental rights in the face of environmental wrongs. Access to justice has been difficult to achieve for victims of environmental catastrophes. The substantive, procedural and evidentiary rules in private environmental actions appear biased in favour of the polluter.[43]

The first point to note is that even a "mediated settlement" is, at best, a fought and won measure based on laws intended to *prevent* the

[41] Grandjean, P. and Landrigan, Philip, 2006, "Developmental Neurotoxicity of Industrial Chemicals", The Lancet, 36; see also Tamburlini, Giogio, 2002, "Children's special Vulnerability to Environmental Health Hazards", in Tamburlini, G. Von Ehrenstein, O., and Bertollini, R. eds, *Children's Health and the Environment; A Review of the Evidence*, EEA Report No.29; Licari, L., Nemer, L. and Tamburlini, G., eds, 2005, *Children's Health and the Environment*, World Health Regional Office for Europe, Copenhagen, Denmark.

[42] McDougall, John, 2012, "Alzheimer's Again Linked to Aluminum", at http://www.rense.com/general137/alum.htm

[43] West, Leigh, 1987, "Mediated Settlement of Environmental Disputes: Grassy Narrows and White Dog Revisited" Vol. 18, *Environmental Law*, 131–150 132.

occurrence of multiple harms. The case involves methyl mercury pollution, contaminating the "English-Wabigoon River system downstream from Dryden, Ontario":

> Two pulp and paper plants in the area, the Dryden Paper Company Ltd., and Dryden Chemicals Ltd., both subsidiaries of Reed Paper Ltd., of England, used mercury cells in sodium chloride electrolysis to produce caustic soda and chlorine.[44]

The harm from mercury pollution is not a new discovery, as alternative technologies had been discovered already in the nineteenth century; also, scientific evidence about the toxic effects of methyl mercury poisoning have been known since the early 1960s.[45] In fact, the Ontario government had sent a team in the Japanese courts.[46]

In this case, similar effects were observed in the Ojibway communities, as the ravages of mercury pollution affect all aspects of the health and the life of the inhabitants. West lists some of the grave problems they encountered:

> ...in the years immediately preceding and following the pollution, the unemployment rate quadrupled from twenty percent to eighty percent...Statistics indicated increases in violence, alcohol-related deaths caused by pneumonia, exposure, and suicide.[47]

In addition, what emerged was "the link between mercury poisoning and the increase in deviant and violent behaviour"[48]:

> ...43.30 tonnes of mercury that was released by Dow Chlor-Alkali plant in Thunder Bay, right next to Fort Williams First Nation, puts into perspective [the Grassy Narrows case]: the Chlor Alkali plant at Dryden (which mercury

[44] West (1987: 133).

[45] West (1987: 133).

[46] See the *Toyama Itai-tai case*, 635 Hanji 17 (Toyama District Court, June 30, 1091); the *Niigata Minimata* case, 642 Hanji (Miigata District Court, Sept. 29, 1971); the *Yokkaichi Asthma* case, 672 Hanji 30(Tsu District Court, Yokkaichi Branch, July 24, 1972); the *Kumamoto Minimata Disease* case, 696 Hanjil 5 Kumamoto District Court, Mar. 20, 1973; reprinted in J. Gresser, K. Fugikura and A. Morishima, *Environmental Law in Japan*, 1981.

[47] West (1987: 135).

[48] ibid.

poisoned Grassy Narrows and White Dog people) only released 10 tonnes of mercury.[49]

In addition, scientist Michael Gilbertson, recently retired from the International Joint Commission, submitted a Ph.D. thesis on December 31, 2006, on the topic of "Injury to Health: A Forensic Audit of the Great Lakes Water Quality Agreement (1972–2005) With Special Reference to Congenital Minimata Disease".

The scientific evidence has been available for years, yet in 1985 when the federal Department of Indian affairs contacted Mr. Justice Emmett Hall (former Supreme Court of Canada Justice), who visited Grassy Narrows and studied a "211 pages legal brief prepared for the Indian Bands by Robert Sharpe, a University of Toronto professor and expert in such litigation",[50] what emerged persuaded him not to recommend going to trial. He believed that the results of the complex and time-consuming litigation would be uncertain, hence that the best interests of the Ojibway's would be served by "a negotiated settlement outside the court system".[51]

One might think that some solution would have been found after the scientific facts had been weighed. Not so, however. The Windsor/Aamjiwnaang FN case is still ongoing, with only "ongoing research regarding the corporate activities" remaining, on the part of the government. The "Grassy Narrows and White Dog" methyl mercury situation was detailed once again by Philippe Grandjean in 2013,[52] and as late as the Spring of 2019, the people of that FN have been demonstrating in front of Toronto, Ontario City Hall, demanding true justice beyond the pittance of support they have received so far. The pulp and paper mill that is poisoning them is still operating at the same location.

[49] Hummel, J.H.W. from links; see also the Madison Declaration on Mercury Pollution, http://www.unbc.ca/assets/media/2007/03march/madisondeclarationonmercurypollutionwithnon-technicalsummary.pdf

[50] West (1987: 136).

[51] ibid., see also Mr. Justice Hall's "Affidavit" before the Supreme Court of Ontario, NO. 14716/77, no. 13.

[52] Grandjean, Philippe, 2013, *Only One Chance (To Develop A Brain)*, Oxford University Press, UK.

5.4 Who Are the Refugees? Escape from Conflict, Starvation and Climate Extremes

> Social justice is a matter of life and death. It affects the way people live, their consequent chance of illness, and their risk of premature death…Within countries there are dramatic differences in health that are closely linked with degrees of social disadvantages….The conditions in which people live and die, in turn, shaped by political, and economic forces.[53]

However, as we have noted in previous sections, economic and social differences are not closer to improvement since 2008, and various groups of refugees, internally displaced persons and others, can only be defined as climate refugees, and their numbers are increasing exponentially, as hundreds of millions of desperate people find themselves with no protection or assistance; in fact these are the people whose very rights are routinely being denied.[54] Yet there are refugees rights in the law, and they should be reviewed briefly, in order to understand what is available legally for their protection.

> Of thirty ways to escape danger, running away is best.
> (Old Chinese Proverb)

With these words, Essam El Hinnawi starts his 1985 monograph on the topic of *Environmental Refugees.* His starting point is the *1972 Stockholm Conference*[55]; he adds, in his discussion, the concept of "Ecodevelopment" coined in this document. Like this concept, which has in recent times morphed into the watered-down notion of "sustainable development", the definition of refugee is not totally "fixed", according to El Hinnawi. He says:

> …environmental refugees are defined as those people who have been forced to leave their traditional habitat, temporarily or permanently, because of a

[53] WHO Commission on Social Determinants of Health, 2008, *Closing the Gap in One Generation*, p. 2.

[54] Passeni, Daniela, Nov.9, 2019, "Climate Refugees could reach 300 million, a population without rights", viewed Nov.21, 2019 at https://global.ilmanifesto.it/climate-refugees-couldreach-JOO-million-a-population-without-rights/

[55] *Stockholm Declaration of the United Nations Conference for the Human Environment*, June 16, 1972, ILM 11141 (El Hinnawi, Essam, 1985), *Environmental Refugees*, United Nations Environmental Program (UNEP), Nairobi, Kenya.

> marked environmental disruption (natural or triggered by people) that jeopardizes their existence and/or seriously affected the quality of their life.[56]

By adding, "In a broad sense, all displaced people can be described as environmental refugees", El Hinnawi places environmental refugees in a category that he views as primary or foundational, rather than simply viewing them as "displaced peoples" instead. These are not accepted as legitimate refuge seekers, according to the 1951 *Convention on Refugees* (CSR), but rather as IDPs, internally displaced persons, not qualified to claim refugee status.

His argument is both correct and appropriate. In turn, he defines "environmental disruption" as:

> ...any physical, chemical and/or biological changes in the ecosystem (or the source base) that render it, temporarily or permanently unsuitable to support human life.[57]

This is the aspect of the issues confronting environmental refugees that is at the heart of this work: when the resource base, the integrity of the lands where a community resides, is destroyed, indeed it can no longer support human life. That has been the argument proposed by my work since 1994.[58] If ecological integrity is central to human health and survival, as well as to the normal functioning of ecosystems, then its absence represents an attack on both health and survival and ecosystem function.

We are all affected in various measures, as I have argued, but the poor and those who live on the land are indeed the most vulnerable. Whether the flight that ensues is temporary or permanent, for most of these refugees, it is indeed the dangerous circumstances in which they find themselves that define their status, even when the circumstances are the result of conflicts or other non-environmental situations. But their numbers are

[56] El Hinnawi (1985: 4).

[57] El Hinnawi (1985), ibid.

[58] Westra, L., 1997, "Terrorism at Oka", in Wellington, A., Greenbaum, and Cragg, W. eds., *Canadian Issues in Environmental Ethics*, Broadview Press, Peterborough, ON; Westra, L., 1998, *Living in Integrity*, Rowman Littlefield, Lanham, MD; Pimentel, Westra and Noss, 2000, *Ecological Integrity: Integrating Environment, Conservation and Health*, Island Press, Washington, DC; Soskolne, Westra et al., 2007, *Sustaining Life on Earth*, Lexington Books, Lanham, MD; Westra, 2006, *Environmental Justice and the Rights of Unborn and Future Generations*, Earthscan, London, UK,; Westra, L., 2007, *Environmental Justice and the Rights of Indigenous Peoples*, Earthscan, London, UK.

often great, and their relocation poses immense problems even in their country of origin. Large migrations, an increasingly common situation today, may well do great damage to the area to which they relocate.[59]

The magnitude of the problem cannot be overstated. A Christian Aid Report predicts that, "given current trends 1 billion people will be forced from their homes between now and 2050",[60] as they add:

> Stranded within their countries and largely ignored by the media, they are the world's forgotten people. The numbered include,
>
> * 50 million people displaced by conflict and extreme human rights abuse. This assumes a rate of displacement of roughly 1 million people a year, which is conservative;
>
> * 50 people displaced by natural disasters. Again, this conservatively assumes that around 1 million people will be displaced this way every year;
>
> * 645 million people displaced by development projects such as dams and mines (at the current rate of 15 million a year);
>
> * 250 million people permanently displaced by climate change related phenomena, such as floods, droughts, famines and hurricanes;
>
> * 5 million people will flee their own countries and will be accepted as refugees.[61]

Already in 1994, the Almeria Statement on Desertification and Migration estimated that "the number of migrants in the world, already at very high levels, nonetheless continues to increase by about 3 million each year".[62] Norman Myers alerted the world to this emerging crisis already in 1993,[63] and in 2005 he "revised his estimate",[64] suggesting up to 200 million as a possible refugee number.[65]

[59] El Hinnawi (1985: 5).

[60] Christian Aid Report, "Human Tide: The Real Migration Crisis", May 2007.

[61] ibid. p. 6.

[62] International Symposium on Desertification and Migration, "The Almeria Statement", Almeria, Spain, 9–11 Feb. 1994.

[63] Myers, N., 1993, "Environmental Refugees in a Globally Warmed World", 43 Bioscience 752.

[64] McAdam, Jane, 2007, "Climate Change Refugees and International Law", NSW Bar Association, 24 Oct.2007:1.

[65] Myers, N. 2005, "Environmental Refugees: An Emergent Security Issue", 13th Economic Forum, Prague, 23–27 May 2005.

However, Sir Nicholas Stern viewed that figure as "conservative", and Myers has since once again revised his estimate.[66] The daunting numbers of ecological refugees, that is, the IDPs that comprise climate and other environmental refugees, as well as those displaced by economic oppression and toxic exposures engendered by "development" and other industrial projects, are grave enough to warrant a thorough re-examination and discussion of present international law.

The main problem with these large migrations lies in the definition of a refugee with its strict limits, according to the *Convention Relating to the Status of Refugees* (adopted on July 28, 1951; into force April 22, 1954):

> *Article 1. Definition of the Term "Refugee"*
>
> (1) ....
>
> (2) As a result of events occurring before 1 January 1951 and owing to well-founded fear of being persecuted for reasons of race, religion, nationality, membership of a particular social group of political opinion, is outside the country of it's nationality and is unable or, owing to such fear, unwilling to avail himself of the protection of that country; or who, not having a nationality and being outside the country of his former habitual residence as a result of such events, is unable, or owing to such fear is unwilling to return to it.

The well-founded fear that is the basis of any refugee's claim appears to be a single individual's sentiment, although, as we shall argue below, it may be viewed from several perspectives, even to apply to whole affected groups, and the Jewish people of Nazi Germany were precisely such a group. Thus, the original intent of these is directed at the plight of single individuals, rather than large migrating groups. But, as El Hinnawi argues, environmental disasters, whether wholly natural (such as earthquakes, droughts and tropical cyclones); partly natural, that is, such that human activities play a significant role in the severity of resulting disasters (such as floods, tsunamis, hurricanes of particular strength); or even those where human activities play a major role, such as melting ice and permafrost in the Artic, rising sea levels elsewhere, are such that invariably large numbers are affected.

All of these disasters affect disproportionately the poorest people in the world, as they have no infrastructure or social services to protect them or

[66] Stern, Sir. Nicholas, 2007, *The Economics of Climate Change; The Stern Review*, Cambridge University Press, Cambridge, UK.

to mitigate the effects of environmental disasters.[67] In addition, land degradation and desertification also render various regions in the South uninhabitable, as does deforestation as these phenomena generate a mass exodus from the affected region.[68] Hence it is easy to understand why other countries and their governments, beyond the areas where the disasters occur, are not eager to open their doors but content themselves with sending aid, at best. This "aid" is often inappropriate, insufficient and hard to distribute to those most at risk.[69]

Often environmental refugees simply attempt to migrate to a different area in their own country, perhaps to the cities. But these are both unprepared and unwilling to receive them and to support their many needs.

5.4.1 Internally Displaced Persons in International Law

> UNDP, in collaboration with IOM, UNICEF and the World Bank, is preparing a handbook on integrating migration into national development strategies. On behalf of the European Commission with other United Nations agencies and IOM, UNDP is managing a programme on knowledge-sharing related to migration and development.[70]

The potentially hopeful presence of the International Organization for Migration (IOM) and Global Forum devoted to the same issue lose their relevance when one considers the main focus of these documents: "development" or, essentially, trade:

> The IOM seeks to engage expatriate communities as partners for development, including through the transfer of knowledge and skills.[71]

Clearly, the drafters do not have in mind the starving millions of IDPs and refugees in sub-Saharan Africa, their deep poverty and the burden of disease, nor do they consider that today's large migrations are, for the most part, the result of trade and "development", both pursued without any

[67] El Hinnawi (1985: 6–20) and Pogge (2009).

[68] Goodland, Robert, 2008, "The World Bank' Financing of Climate Change Damages Integrity", in Westra, L., Bosselmann, K., and Westra, R., eds., *Reconciling Human Existence with Ecological Integrity*, Earthscan, London, UK, 219–244.

[69] El Hinnawi (1985: 23–30).

[70] UNGA, 11 August 2008, A/63/265, p. 11.

[71] ibid. p. 13.

concern for sustainability or the connection between environment and human rights to health.[72]

The basic difference between "migration" and "forced migrations" instead is clearly recognized in the work of refugee of the Refugee Studies Centre at the University of Oxford, together with UNEP, as well researched and analyzed in the latest issue of *Forced Migrations Review*.[73] Achim Steiner, the UN Under-Secretary General and Executive Director (UNEP), says:

> Human migration, forced or otherwise will undoubtedly be one of the most significant consequences of environmental degradation and climate change in decades to come.

Steiner acknowledges that "unsustainable human development", together with sea level rise and the "loss of coastal wetlands", contributes significantly to the millions who will be fleeing by 2080. Rather than focusing on trade advantages, the challenges to human security policies are grave, and "environmental migrants" are recognized now as a separate category, whose numbers have not been acknowledged in law yet, but they are constantly increasing because of "longer term environmental degradation", which aggravates the situation created by the impact of sudden disasters:

> Environmental migrants are understood to be those individuals, communities and societies who choose, or are forced to migrate as a result of damaging environmental and climatic factors.[74]

Climate change impacts will follow three separate avenues: (1) "the effects of warming and drying" will significantly disturb ecosystem services; (2) the increasing presence of catastrophic weather events, such as floods or heavy precipitation, will "generate mass displacement"; and (3) "sea level rise" will permanently destroy the traditional territories of

[72] Gostin, Lawrence, 2008, "Meeting Basic Survival Needs of the World's Least Healthy People: Toward a Framework Convention of Global Health", *Georgetown Law Journal* 9632: 331–392.

[73] *Forced Migration Review*, Issue 31, October 2008, University of Oxford; Marion Couldrey and Maurice Herson, eds., henceforth FMR.

[74] Morton, Andrew Bancour, Philippe, and Laczko, Frank, 2008, "Human Security Police Challenge", FMR, p. 5.

millions.[75] The ongoing development of all these concurrent events demands "proactive intervention"[76]; it also requires the organizational and institutionalization of several areas of research, with a view to promote readiness for mitigation and assistance.[77] These organizations do not have binding force to impose the conclusions they might reach about the necessity of aid and assistance to migrants, let alone the power to be truly "proactive", in the sense of being able to proscribe the hazardous and negligent human activities that produce "overwhelming negative" impacts on peoples and environments and which, in turn, generate mass environmental and climate migrations.

To sum up, the primary responsibility for IDPs rests with their territorial state because of the principles of sovereignty and non-intervention. However, as Goodwin-Gill and McAdam add:

> …the governing principles of sovereignty and non-intervention stand potentially in opposition to often governing principles of international organization, including the commitment to human rights and to international cooperation in the resolution of humanitarian problems.[78]

The major problem is that "neither the UNHCR nor any other UN agency has any legal authority 'to protect' persons within their own country".[79] Hence the protection of IDPs, although a "broad consensus" exists in principle on the need to protect,[80] remains unsolved, at least at the international level. The question on the ground remains almost exclusively regional and political, although for certain specific circumstances the internally displaced may be considered in international law, given the ongoing failures of states to provide protection:

[75] ibid. p. 6.

[76] ibid.

[77] The Climate Change Environment and Migration Alliance (CCEMA), was established in April 2008 in Munich, by the UNU and the IOM, UNEP, and the Munich Re Foundation (MRF) with the goal of ensuring both preparedness and collaboration.

[78] Goodwin-Gill, Guy S., and McAdam, Jane, 2007, *The Refugee in International Law* 3rd ed., Oxford University Press, Oxford, UK, p. 33.

[79] ibid. p.34; see also UNGA res. 60/168, "Protection and Assistance to Internally Displaces Persons", 16 Dec.2005 (adopted without a vote).

[80] ibid. p. 48.

In each case however, there is a clear gap between what may be called *functional* responsibilities and expectations, on the one hand, and the legal obligations of States, on the other hand.[81]

The presence of such "gaps" does not appear to be purely an oversight, as states have consistently avoided the drafting of a new, binding convention that might solidify and render mandatory the rights of IDPs in relation to their state of origin. Since 1990, however, the legal concept of "complementary protection" has had a history dating back to the League of Nations,[82] as the formal "refugee" definition was found to be increasingly incomplete. Complementary protection was intended as a legal term to be distinguished from the protection "granted solely on humanitarian grounds", based on humanitarian and human rights law.[83] However neither human rights nor humanitarian law applies to all who may need protection: for instance, the norms of humanitarian law are applicable only during an armed conflict, and other forms of "low-intensity conflicts" do not qualify.[84] Human rights law also needs the clarification as the protection of life is certainly present within it, as are "free movement and access to international assistance", as well as "the right not to be arbitrarily displaced".[85]

In addition, the right not to be arbitrarily displaced is not explicitly formulated in any human rights instrument, except the ILO Convention No.169, but that document refers exclusively to Indigenous and tribal peoples. The Universal Declaration of Human Rights includes implicitly applicable language, and the Human Rights Committee in 1999 stated that,

[81] ibid. 47.

[82] Marrus, M.R., 2002, *The Unwanted: European Refugees from the first World War Through the Cold War* 2nd ed.

[83] Goodwin-Gill and McAdam, 2007:286; see also Executive Committee, Standing Committee, 18th Meeting, "Complementary Forms of Protection: Their Nature and relationship to the International Refugee Protection Regime", UN Doc.EC/50/SC/CRP.18 (9 June 2000) esp. para 4 and 5.

[84] Phuong, Catherine, 2004, *The International Protection of Internally Displaced Persons*, Cambridge University Press, Cambridge, UK, 48–49.

[85] Phuong, 2004: 51; see also Stavropoulos, M., 1994, "The Right Not to be Displaced", 9 *American University Journal of International Law and Policy*, 689.

> ...the freedom to choose one's residence[86] includes protection against all forms of forced internal displacement.[87]

The only instrument that defines specifically the rights of IDPs and states and others' obligations is the Guiding Principles mentioned above, but these Principles do not represent a binding obligation and most of the other international instruments from which the protection of IDPs might be inferred, including both "ratification gaps" and "consensus gaps"[88], hence we can conclude that, at this time, no binding instrument exists for the protection of all IDPs, whereas perhaps Indigenous peoples may fare slightly better.

Nevertheless the problem of IDP protection remains a highly significant one, not only because of the presence of climate change and the pressures of globalization that ensure ever-increasing numbers of migrants who are presently unprotected but also because the problem it represents reflects the culmination of the modern tension between sovereignty and human rights problems.[89] For this reason, primarily, refugee law, that is, the 1951 CSR, is not only a flawed instrument as such, as we shall see below it cannot properly protect even all its own defined "clients", but it is also less than useful to accommodate the growing new phenomenon, IDPs. And this is true even though most of the categories of "persecution" on which it is founded would lend themselves well to a possible extension to IDPs, at least in principle.

At any rate, solutions for both categories of affected persons, refugees and IDPs would benefit more from the immediate adoption of preventative measures to protect them "at source", that is, long before the circumstances of their existence deteriorate to the point at which flight is the only option for survival.

[86] Article 12(1) of the ICCPR.

[87] Human Rights Committee "Freedom of Movement", (Article 12) CCPR/C/21/Rev.1/Add.9, CCPR General Comment 27, 2 November 1999; see also discussion in Phuong, 2004: 51.

[88] Phuong (2004: 49–50).

[89] Anghie (2006); see more specifically Dacyl, J., 1996, "Sovereignty Versus Human Rights: From Past Discourses to Contemporary Dilemmas", 9 *Journal of Refugee Studies*, 136.

5.5 People Without Rights: Whose Responsibility?

A Tuvalu family has been granted New Zealand residence after claiming it would be affected by climate change if it returned home. It is the first successful application for residence on humanitarian grounds in which climate change has featured, but the Immigration and Protection Tribunal said the family had strong ties to New Zealand.[90]

The Tuvalu family succeeded, whereas the same year a family from Kiribati's request was rejected. Kiribati, like Tuvalu, is slated to be so on under water, to disappear beneath the sea as the effects of climate change persist. The father who lodged the request said, "When Kiribati disappears, we're going to Die with our Kids". As we saw, human rights for refugees of extreme situations are clear and well-established. Climate change is the most obvious and visible cause of necessary flights from the horrors encountered by those who live in certain areas.

Hunger, as we argued, is also the necessary result of climate change, coupled with the corporate practices of land grabbing, so that you do not see two completely separate causes, but two complementary problems sharing the same etiology. Also, many other reasons based on human rights also exist and should be added to the totality of major issues that continue to be ignored when refugees are at the border of today's states.

5.6 Concluding Thoughts

A domino-like cascade of melting ice, warming seas, shifting currents and dying forests could tilt the Earth into a hot-house state beyond which human efforts to reduce emissions will be increasingly futile, a group of leading climate scientists has warned. The grim prospect is sketched out in a journal paper that considers the consequences of 10 climate change processes, including the release of methane trapped in Siberian permafrost, and the impact of melting ice in Greenland on the Antarctic.[91]

[90] Maas Amy, 2014, "Tuvalu Climate Change Family wins NZ Residence Appeal", *New Zealand Herald*, 3 August 2014.

[91] Watts, Jonas, 7 August 2018, "Domino Effects of climate events could move Earth into a hothouse state", viewed Nov.24, 2019 at https://www.theguardian.com/environment/2018/aug/06/domino-effect-of-climate-evants-could-push-earth-into-a-hothouse-state, essay Published in *Proceedings of the National Academy of Sciences*.

The article continues discovering yet another dangerous "feedback loop" as "increased rainfall-a symptom of climate change in some regions-is making it harder for forest soils to trap greenhouse gasses such as methane".[92]

Here we see further confirmation of our thesis, that is, the inescapable connection between climate change and starvation, where methane gas has been shown to be a determinant part of the climate change situation, as it adds its presence beyond industrial battle operations. *Bioscience* published as far back as 1988 an insightful paper on Brazil, entitled "Environmental Degradation in the Pantanal Ecosystem" (in Brazil, the world's largest wetland is being threatened by human activities[93]):

> The Pantanal, with its extraordinary diversity and Abundance of wildlife, is a threatened region. Deforestation, expanding agriculture…the pollution of the water with herbicides, pesticides, and by-products of fuel alcohol production have caused the progressive deterioration of the natural environment, placing at risk of Brazil's most important ecosystems.[94]

Further, this article adds that "no research has been conducted on the fates of herbicides and pesticides in the Pantanal", as they mention 2,4-D and 2,4,5-T and even the use of Agent Orange in the area.[95] It would seem there is no excuse for policymakers and governments to allow the continuation of disastrous practices that have been clearly known for twenty years.

The question that arises is who then can defend the forests of Brazil? Who can actively engage against climate change and starvation? All countries thus far have failed spectacularly as they catered to corporate interests rather than to science. It appears that, practically today, the only defenders of the area appear to be the Indigenous peoples, with bows and arrows as their only weapons. The European Union has just declared a climate emergency, as Ursula von der Leyen has named Italy's Venice, now

[92] PNAS published paper Aug 14, 2008 PNAS (National Academy of Science), 115()3) 8252–8259, https://doi.org/10.1073/pnas18101411151; "Trajectories of the Earth System in the Anthropocene".

[93] Alho, Clebo, J.R., Lacher, Thomas E., Jr. and Gonalves, Humberto C., 1988, "Environmental Degradation in the Pantanal Ecosystem, in Brazil, the world's largest wetland is being threatened by human activities", *Bioscience*, Vol No.3, pp. 164–171.

[94] Alho et al. (1988: 164).

[95] ibid., p. 167.

partially submerged and with all its beautiful art and monuments at grave risk, as its "poster child" of the movement she proclaims.

The American continent, in contrast, remains silent limiting itself to political battles and building walls, behind which to hide, a futile enterprise. The Indigenous peoples of the world understand all this better than the politicians do. A short while ago, a Western man with religious pretensions attempted to civilize a community, living in one of the Indian Ocean's biodiversity paradise, its ecointegrity untouched by the local group who lived their lives in harmony with nature. The Western man's first visit did not end well; the group was clearly not interested, as they pointed their bows and arrows his way.

On his second and final visit, they followed through with their threat, and they buried him in a shallow grave close to the seashore.[96]

I presented this case as part of a summary of my (then) forthcoming work (Westra, L, 2019, *From Principles to Norms: The Development of International Law*, Nova Science Publishers, New York). We have been convinced that developing our "civilization" was a good thing, beyond discussion. I wanted to raise a question: is it right, even permissible, to kill for basic, foundational principles? The island where those Indigenous peoples lived, and still live, is a biodiversity paradise; was it right on their part to defend their cultural and ecological integrity at all costs? The closest police took that position too, as there was no prosecution involved, although it remained clear who did the killing. Those to whom I posed the question agreed that it was a case of justifiable self-defense.

In the Summer of 2019, we see the ongoing attacks on Indigenous peoples, intent on living undisturbed as well: in Brazil, in the Amazon Forest. Yet the leader of an Indigenous community was shot dead as he attempted to resist the invading loggers.[97]

The Indian Ocean's islanders were the protectors of their cultural and ecological integrity, I argued, hence had the right of self-defense. Similarly, and even more strongly, the defenders of the Amazon Forest have the right of self-defense, the same cultural and ecological rights of the other

[96] Conroy, J. Oliver, Feb.3, 2019, "The life and death of John Chau, the man who tried to convert his killers", *The Guardian*, available at https://theguardian.com.world/2019/feb/3/John-Chau-christian-death-sentilese

[97] Sam Cowie, Sao Paulo, 2 Nov. 2019, "Brazilian Forest Guardian killed by illegal loggers in ambush" (Paul Paulina Guajajara was killed by armed loggers in the Araniboia region in Maranhao), viewed Nov.3, 2019 at https://www.theguardian.com/world/2019/nov/02/brazilian-forest-guardian-killed-by-illegal-loggersin-ambush

Indigenous peoples' community. But, in addition, they could also claim that—at the same time—they were defending the rights of survival of humanity as a whole, through the protection of their ecological integrity.

In the light of this work, perhaps another question is in order, do we all have the right of self-defense in the light of what states and their corporate supporters are doing to our physical integrity and that of the Earth, perhaps not by outright killing as the Guardians of the Forest are prepared to do, but surely by invoking criminal prosecution for the protection of the Amazon Forest and ultimately of ourselves? In the next and final chapter, we will describe what should have been a better path, although it too would lead to the same conclusion.

References

421 US 1, 96S.ct. 612, 46L

424 U.S. 1 (1976)

A/HRC/34/18, para. 86

Alexander Zaitchik, July 6, 2019, "Rain Forest on Fire (On the Front Lines of Bolsonaro's War on the Amazon, Brazil's Forest Communities Fight Against Climate Catastrophe)", https://theintercept.com/2019/07/06/brazil-amazon-rainforest-indigenousconservation-agribusiness-ranching/ viewed on Aug.26, 2019

Anghie, Thomas, 2006, *Imperialism, Sovereignty, and International Law*, Cambridge University Press, UK

Barlow, Maude and Clarke, Tony, 2003, *Blue Gold* (The Battle Against Corporate Theft of the World's Water), "Losing the Great Lakes" pp. 35–6, McLelland and Stewart, Toronto, Ont.

Bruch, Mark, Gov. Snyder moves to come up with $12 million to switch Flint's water back to Detroit's supply, viewed Aug.2 at http://michiganradio.org/post/gov-snyder-moves-come-12million-switch-flints-water-back-detroits-supply, Michigan Radio Oct.8, 2015

Bullard, R.D., 1994, *Dumping in Dixie*, Westview Press, Boulder, CO

Dellapenna, Joseph, 2017, "Flint's Water crisis: Profitability, Cost-effectiveness, and Depriving People of water", in press

Dockrill, Peter, *Science Alert*, June 26, 2019, viewed Sept. 22, 2019 at www.businesainsider.com/climate-apartheid-united-nations-report2019-6?IR=T

Fonger, Ron, Jan. 2, 2015, "Lead leaches into very corrosive Flint drinking water, viewed Aug. 2, 2016 at www.mlive.com/news/flintindex.ssf/2015/09/new_testing_snows_flint_water.html, Mlive Sept.2, 2105; "Flint returning to Detroit water amid lead concerns"

Fonger, Ron, June 4, 2014, "City warns of potential health risks after Flint Water tests revealed too much disinfection byproduct", *Flint Journal*, Mlive Media Book

Grandjean, P., 2013, *Only One Chance*, Oxford University Press, Oxford, Chapter 3

Hanna-Attisha, Mona, Lachance Jenny Sadler, Richard Casey, Champney Schrepp, Allison, 2015-12-21, "Elevated Blood Lead Levels in Children associated with the Flint Water Crisis: A Spatial Analysis of Risk and Public Health Response", *American Journal of Public Heath*, 106(2), 283–290

Herbertson, Kirk, Thompson, Rim and Goodland, Robert, 2010, *A Roadmap for Integrating Human Rights into the World Bank Group*, Washington, DC, World Resources Institute

Howard, Philip H. 2009, "Consolidation in North American Organic Food Processing Sectors 1997-2007", International Journal of Sociology, Agriculture and Food, 16(1), 13–30

Howard, Philip, Aug.4, 2012 at http://www.msu.edu/-howardp/organicindustry.html

ILC (2011) "Tirana Declaration" Securing Land access for in times of intensified natural resources competition", International Land Coalition, viewed July 12 at http://www.Lnadcoalition.org/about-us/aom/2011/tirana-declaration

Landrigan, Philip and Ruth Etzel, 2013, *Textbook Children's Environmental Health*, Oxford University Press, New York

Lanphear, Bruce P., 2013, "Lead", in Landrigan, P. and Etzel, Ruth, op.cit., pp.262–264

Liberti, Stefano, 2013, *Land Grabbing: Journeys in New Colonialism*, Verso Publishing, London, pp.144, ff.

Lin, Jeremy C.F., Rutter, Jean, Park, Hayoun, Jan.21, 2016, "Events that led to Flint's Water Crisis", *New York Times*

Margulis and Porter, 2014:73

Martin, Melanie J., "ALEC Les Corporations Craft Environmental Policy", 29 July 2012

Mattei, Ugo and Nader, Laura, 2009, *Plunder*, Blackwell Publishing, p. 181

McMichael, Philip, 2014, "Land Grabbing as Security Mercantilism in International Relations", in Matias Margulis, Nora McKeon and Saturnino M. Borras eds., *Land Grabbing and Global Governance*, Routledge, pp.47–64, p.49

Oxfam Annual Report, 2011

Pearce, 2012, *The Land Grabbers: The New Fight Over Who Owns the Earth*, pp.202–203

Peluso, N.L. and Lund, c., 2014, "New Frontiers of Land Control", The Journal of Peasant Studies, 38(4), pp.667–681, pp.673–4

Pogge, T., 2009, "Aligned Global Justice & Ecology" in Westra, Bosselmann & R. Westra eds., *Reconciling Human Existence with Ecological Integrity*, Earthscan, UK, pp.146–158

Shiva, Vandana, 2002 *Water Wars* (Privatization, Pollution and Profit), Between the Lines, Toronto, Ontario

Shiva, Vandana, 2016 "Make Monsanto Pay", *Commondreams*, Feb.24, viewed August 19, 2019 at http://www.commondreams/org/views/2016/02/24/make-monsanto-pay

Tamburlini, G., 2002, "Children's Special Vulnerability to Environmental Health Hazards", in G. Tamburlini, O. von Ehrenfeld and R. Bertollini, 2002, *Children's Health and the Environment: A Review of the Evidence*, EEA Report No. 290 WHO, Geneva, Switzerland

UNGA/HR /34/48

United Nations Office of the High Commissioner on Human Rights, Statement by Prof. Philip Alston, UN Special Rapporteur on Extreme Poverty and Human Rights, Malaysia, 13–23 August 2019

Venton, Peter, 2020, The Political Economy of Managing Without Growth, in press

Westra, Laura, 2017, *On Hunger*, Brown Press, FLA

Westra, Laura, 2018, *Environmental Disasters and Land Grabs as Crimes Against Humanity*, Nova Science Publishers, New York

Westra, Laura, 2019, *From Principles to Norms: The Development of International Law*, Nova Science Publishers, New York

WHO, 2012, *State of the Science of Endocrine Disrupting Chemicals 2012: World Health Organization Summary for Decision Makers*, World Health Organization, Geneva

Winston, Samuel, October 7, 2015, "How the Flint water Crisis Emerged", *Flint Journal*, p.2

CHAPTER 6

The Principle of Integrity Revisited

Abstract We have now reached a proposed better approach, an approach that has been used through several decades and has helped to produce some of the improvements we have noted and raised awareness on most of the issues, as well as help to disseminate the necessary information. Ecological integrity was acknowledged following the US clean Water Act (1972) and the Great Lakes Water Quality Agreement (1987, rat.1988). An Index of Biotic Integrity (IBI) designed by a Seattle biologist has been used since then to reach policy decisions regarding water-related issues in the US. Ecological integrity was embraced by a group of scientists, ethicists and eventually lawyers. However the general term ecological integrity was also followed by the acknowledgment that single organisms also possessed an integrity that could be breached. Thus GMOs were indicted together with the other toxic substances employed by agribusiness. The chapter traces the growth of what is now a movement, active at many international venues, primarily in Europe but also internationally at the United Nations. Supporters of integrity have not reached a solution, but they have been, and they are part of the fight.

Keywords Ecological integrity • Biological integrity • Index of Biotic Integrity • GMOs

L. Westra, *Climate Change and Starvation*, Palgrave Insights into Apocalypse Economics,
https://doi.org/10.1007/978-3-030-42124-3_6

6.1 Introduction: From Aldo Leopold to the Principle of Integrity

Aldo Leopold: "A thing is right if it tends to preserve the ecological stability, integrity and beauty of the biotic community and wrong if it tends otherwise".

The Principle of Integrity:

(1) The first moral principle is that nothing can be moral that is in conflict with the physical realities of our existence, or cannot be seen to fit within the natural laws of our environment.

(1a) Act so that your actions will fit first and minimally within universal natural laws.

(1b) Act so that you manifest respect and understanding acceptance of all-natural processes and laws (although self-defence is acceptable).[1]

In 1994, the main discussion hinged on the values underpinning ecological integrity. As the argument at that point was based primarily on philosophy, my earlier discipline, but also somewhat defensively on science, the biological integrity of waters, but also on the ecosystem approach, which was not generally acknowledged to be valid; thus the science of integrity had to fight for recognition for many years. It was based on the 1972 Clean Water Act (US) and on the parallel work of the Great Lakes Water Quality Agreement and the work of James Karr, a biologist at the University of Seattle.

Karr did not view ecological and biological integrity as separable; the water organisms that were his main concern actually served as indicators of the integrity of the water wherein they lived. The presence of healthy specimen, from bacteria to salmon, meant that the water of the river or lake had integrity. The period from 1992 to 2000 was used in concerted efforts at each yearly meeting, to understand the value of ecological integrity and find ways to measure its presence. Eventually in 2000, a multidisciplinary work compiled the most significant discussion and definition of integrity for our group.

At that point the emphasis on the integrity of single organisms was confirmed, but that strand was eventually developed in the years after 2000. However, as early as 1988, I published in a journal of social science an article against transgenics, which were the precursors of GMOs, and

[1] Westra, Laura, 1994, *The Principle of Integrity*, Rowman Littlefield, Lanham, MD.

some of us continued to pursue that line, leading to questioning the content and safety of the food that was commonly consumed, as well as the policies that were intended to protect that food's quality for consumption.

In the previous chapter, we noted the words of Pope Francis, his insistence on treating corporate interference with natural systems as a crime, in fact as "ecological sin". But it is important to note that many decades ago, other religious groups took a very similar stance, based like that of the Pope, on both science and religion, the Dutch Reformed and especially the Mennonite families of Kitchener, Ontario, in Canada. They jointly published a journal with the suggestive title of *Earthkeeping* (A Quarterly on Faith and Agriculture), and in this chapter, we will discuss their findings and their position. Some of these can be viewed driving west from Toronto, as they drive their horse and buggies within sight of the only major highway leading from Toronto, Ontario, to Detroit, Michigan.

Thus we will review the history of our enterprise, but, beyond reviewing that material on integrity, it will be important to also review the strategies we have followed, especially with regard to the law, both domestic and international, to the diffusion of information through various channels and its indirect impact on climate change, as well as its even more significant impact, direct this time on starvation.

The impact on legal regimes is increasingly visible, as the information and knowledge related to it has contributed to various movements and even to the growing commitment to halt/reverse or at least modify the presence of climate change, at least in the European countries. That is the most important aspect of what is now known as the Global Ecological Integrity Group, which has even clearly incorporated two separate but intimately connected aspects, as was formally agreed upon in 2017, at the time of our 25th anniversary meeting in Windsor, Ontario, the location where it all started.

Our yearly meeting would provide the foundational theoretical basis, encompassing all the present discipline present now in our group, perhaps with a preponderance on legal scholars, but many of the younger, more active legal representatives would then work with other related groups and associations, intent on the diffusion of our message from the UN to a number of international related meetings, to maximize the impact of what might be the advances and discoveries from each meeting.

6.2 The Early Quest for Values Underpinning Integrity

From the start, the emphasis was on understanding and defending integrity. Originally accepted by some scientists as a truly scientific concept, justified for decades after its first appearance in the wording of the 1972 US Clean water Act, then later becoming foundational to the *Great Lakes Water Quality Agreement* of 1978 (ratified in 1988). Details of both documents were discussed in earlier chapters. At that time, the ecosystem approach, now widely recognized, was viewed with a great deal of suspicion in some scientific circles, as not precise enough for a scientific approach. Those who held that position fought with any means available to discredit those of us who knew the importance of that starting point. Hence it became extremely important to support integrity from a philosophical point of view, as from that perspective it seemed to be entirely defensible.

The analysis of several different aspects of integrity discloses a number ofvalues that can be attributed to it. They are primarily non-anthropocentric, although most have further anthropocentric applications, not constitutive of the value of integrity as such. These values can be listed under ten headings.

Integrity's value may be based on the following intrinsic aspects:

1. It is a universal value, global in a sense different from both peace and health, the only other values that may compare in relevance with it.
2. It is a revolutionary concept, a catalyst for the necessity of rethinking both national basic beliefs about just human interaction and human rights and accepted standards of previous international interaction.
3. It emphasizes the meaning of its paradigm case, organic unity and the value of freedom in its two senses, negative and positive, in that regard. Negative freedom requires no interference with the organism's biological identity; positive freedom requires that the organism's conditions be such that it may continue in its capacity to develop and change while retaining identity.
4. It includes the value of health in a non-anthropocentric sense.
5. It indicates and supports the value of the whole, while correspondingly emphasizing the reduced role and status of individuals

including humankind. This emphasis is only applied to biological interaction, not to social or cultural activities.
6. It supports the value of harmony, including that between individuals and whole, and structure and function, once again, only in the biological sense indicated in 5 above.
7. It encompasses the value of sustainability, hence stability. In fact, it is absolutely necessary in some proportion, for the sustainability of all other parts of the landscape.
8. It emphasizes the value of life/existence as such (i.e., for all individuals and wholes).
9. It shows the congruence of morality, science/empirical reality and metaphysics, in integrity as an environmental value.

Not everyone will accept each of these separate, interrelated facets of the value of integrity. Nor do I claim that only these are all its possible valuable aspects. Nevertheless, even some of these aspects manifest the existence of a cluster of values that can be discovered as implicit in the concept of integrity. And even accepting some of these might be sufficient, in order to support the principle of obligation arising from integrity.

Those who opposed the ecosystem approach claimed to be the true representatives of the scientific method and refused to debate with others who disagreed, in contrast to the defenders of integrity who accepted different approaches and methodologies, which were eventually conducive to the drafting of a thorough, multidisciplinary definition of ecological integrity in 2000.

However, several years earlier, there were groups that combined a scientific approach to agriculture to that of the protection of the Earth with a strong religious belief: they were the Dutch Reformed and the Mennonites in the Kitchener/Waterloo area of Ontario, in Canada. They could be seen driving their horse-and-buggy conveyances from the main highway, going west from Toronto, Ontario, to Detroit, Michigan.

They published a journal, titled *Earthkeeping: A Quarterly on Faith and Agriculture*.[2] What makes this journal highly significant, in fact unique, is that not only the language used in several of the articles, or the letters from members, but also its approach to science in regard to the cultivation

[2] Published by the Christian Farmers Federation of Alberta and Ontario. The most evocative is an issue I was able to retrieve, Vol.6, No.3, March, 1991.

of the Earth is quite close to the recent work of Pope Francis own work, such as his 2015 Encyclical entitled *Laudato Si'*.[3]

The 1991 sample of that early publication is a true treasure trove of information. It includes a wonderful short article entitled "The Organic Agricultural Movement: A Historical Perspective", by Dennis Vriend[4]:

> modern organic practices have their roots in the marriage of two sciences chemistry and biology…on another front, a second science was emerging. Louis Pasteur's microscope revealed a whole new world of microbes, plants and animals. Suddenly soil was no longer simply *terra firma*, but living earth made up of living organisms.

Apparently Vriend noticed that some critics ridiculed these farmers' efforts "to uphold the integrity of creation", as inefficient practices, because they are based on rotation and legume plantings, rather than depending on "artificial fertilizers".[5]

Another article by Valentine Yutzy, "Cooperating with Creation", focuses on soils, as he remarks "soils that are biologically alive have less erosion problems".[6] Others proclaim to be opposed to artificial fertilizers and pesticides. Brian Vanderberg, a Dutch Reformed Christian farmer, remarks that "Each day anew we are reminded that God's creation is complex and integrated", as, he adds, "When God saw all he had made, it was very good".[7]

Thus, the connection between integrity and food production was acknowledged long before the recent "awakening" present at least in Europe. It is worth noting that the "ridiculing" effects felt by the Christian and Mennonite farmers who subscribed to the defense of integrity was also felt by many who were fearful, or at least seriously concerned, about practices current at the time and the health risks involved. A social scientist, Edith Efron, published what—at the time—was an influential book.[8]

[3] Pope Francis, 24 May 2015, "Encyclical Letter, Laudato Si' of the Holy Father Francis on Care for Our Common Home", available at http://w2.vatican.va/content/francesco/en/encyclicals/documents/papa-francesco_20150524_encyclica-laudato-si.html

[4] op/cit. *Earthkeeping*, 1991, pp. 14–15.

[5] ibid.

[6] op. cit. p. 11.

[7] Genesis 1:31a.

[8] Efron, Edith, 1984, *The Apocalyptics1 Politics, Science and the Big Cancer Lie*, Simon and Schuster, New York.

Her argument was that those who fear environmental hazards (the example given was carcinogenic exposures) do not do so because of rational fears, but because of their "fantasies" about "industrial mass murder": they all intend to blame industry for non-existent cancer risks. The title of Efron's book is particularly interesting, in contrast with the title of the present work: and the innocence of corporate products and practices is no longer defensible or even considered a laughing matter.

6.2.1 *The Concept of Integrity Defines Its Value*

At any rate, the general hostility against the basic tenets in support of integrity, that is, the ecosystem approach, forced a highly detailed defensive approach when addressing the ecosystemic aspects of integrity.

Because the "principle of integrity" is based primarily on ecosystem integrity and the closely related concept of global "ecological integrity", it is important that this sense of integrity is shown to be consistent with the previous one and that it can be shown to represent a stronger basis for supporting ethical obligation.

Now a philosopher may speak in general terms of 'the earth" or the environment, but a scientist must be more specific. Even if she avoids the tight confines of the "reductionistic" approach (both as a methodology and as an ontology), factual descriptions and hypotheses meant to fit physical reality must be detailed in order to be useful. Consider the definitions of the useful term's "biome" and "ecosystem":

> A biome is the largest land community unit which ecologists find convenient to recognize. It refers to a biogeographical region or formation; a major regional ecological community characterized by distinctive life-forms and principal plant and animal species.
>
> Ecosystems, however, are smaller units than biomes: an ecosystem comprises a natural community and its abiotic environment together as a functional system of complementary relationships in which there is transfer and circulation of energy and matter.[9]

Jeffrey McNeely suggests that the biome is the best approach within the biosphere or the entire global ecosystem for defining a community capable of supporting living organisms. It offers several advantages, such as (1)

[9] Odum, E. P 1993, "The Ecosystem", *Ecology*, 2nd ed., Sinaver Association, Sunderland MA.

recognition of the necessity of development based on natural systems of renewable resources, in turn based on a broad perspective; (2) identification of major common "ecological relationships, processes, potentialities, and vulnerabilities"; (3) identification of major common problems (thus fostering the quest for common solutions); (4) stimulation of interdisciplinary, comparative and international studies and applications to common problems; and (5) identification of "priorities of conservation", based on scientific grounds, "regardless of the location and national frontiers" (Ibid.). The concept of "ecosystem" provides a unit whose smaller scale allows more ease of handling than a larger unit such as the biome. Yet when philosophers use the term "ecosystem", they might mean biosphere or the environment in a more general sense. The objects of their focus are the universal features of all ecosystems. This seems justified given the universality of the principles they want to reach. What do ecosystems have in common? Ecologist Eugene Odum, for instance, cites the dual role of ecosystems as not only "supply depot(s)" but the "home(s)" in which we must live. This "home" supplies us with services arising from "non-productive landscapes": we need a "balanced CO_2O_2 atmosphere.... the climatic buffer provided by oceans amid masses of vegetation...".[10]

It is worth noting that the "service" approach is clearly anthropocentric and utilitarian. Rather than conceiving of healthy, diverse ecosystems as commercially valuable monocultures, we see them as contributing "to the long-term creativity of nature – an area of great biotic diversity creates more diversity through time". Various scientifically observable processes, creating a more "heterogeneous environment, result in and "explain why total diversity is self-augmenting in ecological time", as well as "in evolutionary time".

On the other hand, its reverse has the reverse effect, hence the far-reaching results of disintegrity: "Losses in diversity beget further losses." B.G. Norton adds:

> If a species goes extinct, other species that interact with and depend upon it are threatened. A well-known biologist has asserted that for every plant species that becomes extinct, fifteen animal species can be expected to follow.[11]

[10] ibid.

[11] Norton, B.G., 1986, Why Preserve Natural Variety? Princeton University Press, Princeton, NJ, pp. 49–60.

Holmes Rolston says:

> We want to value the lush life that ecosystems maintain in their diversity, unity, dynamic stability, spontaneity; the dialectic of environmental resistance and conductance; the generating life forces.[12]

This represents the basis for the "principle of integrity" we wish to defend. Since "unity, completeness and value" are among the various dimension of integrity, it goes beyond the concept of health.

The values of "health" and "integrity" appear to be independent, although often connected to one another. Further, even in the ecosystemic sense, the concept of "integrity" itself may be said to have two aspects: (1) structural integrity (I_1) and (2) functional integrity (I_2). Functional integrity might be easily associated with stability and with ecosystem health.

In essence, our aim is to discover and defend a value strong enough to ground an obligation. For this purpose, "structural integrity" might be more significant than functional integrity since it facilitates the differentiation of "integrity" from notions of health, continued existence or stability. Structural integrity may then form the basis for both the "dignity" of the environment and the obligation of respect in regard to it. This appears to imply that "structural integrity" might be relative to some kind of phenomenon that is both cultural and natural, even if no longer pristine. According to this model, one would need a further distinction between "pristine structural integrity" and "transformed structural integrity", and an example of the latter might be a restored ecosystem, supporting sustainable development. Yet "structures" may be understood as "slow processes or functions" and processes (functions) as "rapidly changing structures". Robert Rosen discusses the concept of "function" in a chapter entitled "Entailment Without States: Relational Biology", in his work *Life Itself*. He speaks of "heterogenous systems", wherein "one part looks different, or behaves differently from another part". He continues:

> If we leave the system alone, some autonomous behavior will ensue. On the other hand, we can ask a question like; if we were to remove, or change, one of these distinguishable parts, what would be the effect of that behavior? This is a pregnant question. It involves a new element, nor merely

[12] Rolson, Homes III, 1988, *Environmental Ethics*, Temple University Press, Philadelphia.

> observation, but willful active intervention. The result of that intervention, is in effect, the creation of a new system, which can be regarded as a kind of perturbation or mutilation of the original one.[13]

The natural, "autonomous" behavior of a system is contrasted with that of a system after "willful, active intervention". The intervention's result is another system, a "mutilation of the original one". It is not necessary to totally qualify, quantify and define the former, to be able to say the latter is "mutilated", thus at best a healthy ecosystem and, at worst, a measurably degraded one. "Mutilation" is a pejorative term, although the ensuing ecosystem might be supported/manipulated for desirable human goals. Integrity refers instead to the ecosystem in its unmutilated condition.

To seek a possible answer to these difficulties, it is worth returning to the science of ecology and its main current paradigms. The ecosystem approach is by no means the only available "window" through which these can be studied. Indeed, the concept of an ecosystem itself is fraught with fundamental ambiguities. At any rate one major approach views individual organisms, aggregates and communities as the main basic units of study, thus reducing ecosystems to "relatively constant backgrounds". Another views ecosystems (however defined) as primary and considers the relevance of succession and change first, thus reducing individuals and communities to the role of members/parts of the ecosystems. Both approaches are valid and in fact necessary for better comprehension of both composition and functioning of ecosystems.

The latter could be defined as "any system that…(is) open to the flow of energy and matter and (contains) at least one living entity", yet this offers no answers to persistent questions of time and scale, without which ecosystems cannot be understood properly. Ecosystems can also be defined as "the smallest units that can sustain life in isolation from all but atmospheric surroundings", although there is no "simple method for specifying the size of an ecosystem unless recognizable boundaries exist".[14]

Hence, those who are accustomed to more easily identifiable objects of study, such as organisms or communities (although the latter may not be totally precise themselves), tend to downplay the ecosystem approach, as

[13] Rosen, Robert, 1991, *Life Itself, A Comprehensive Inquiry into the Origin, Nature of Life*, Columbia University Press, p. 116.

[14] O'Neill, Robert, DeAngelis, D. et al., 1986, The Hieratical Concept of Ecosystems, Princeton University Press, Princeton.

based on entities too ephemeral and imprecise to support a scientific theory. Aside from such debates, it is clear that different scientific projects may well require different approaches. On the other hand, the concept of "integrity" may well provide an "umbrella" under which both aspects, that is, that of function and that of structure, retain their relevance. Integrity takes into consideration both organisms and communities within an ecosystem, as both additions and subtractions of species. The addition of species alien to the ecosystem under consideration is viewed as inappropriate—at least prima facie—from that point of view, and the introduction of exotics appears even less recommendable.

The functions and processes taking place within an ecosystem, therefore, needed to be understood through careful consideration of its structure, and the organisms and species that inhabit and form it also cannot be understood in isolation (ibid.) The same is true of human organisms: a human being's physiology can be studied as such, but stresses and disturbances, that is, its pathologies, require additional investigation of its environment. Viruses and bacteria, for instance, interact with and within the human system, and so do food, water and air that sustain its life, as well as the toxins, pollutants and man-made chemicals that might be ingested, absorbed through outgassing of chemicals in its surroundings.

In this sense, some of the problems concerning the precise definition and boundaries of ecosystems may need to be rethought as requiring too limiting and reductionist a view. Taking the example of the human organism once again, the same question can be raised: what are our "limits"? Is it simply our skin that limits the individual a scientist or doctor needs to understand? What of genetic background, medical history, lifestyle present surroundings and possible exposures, as well as emotional attachments or isolation? A doctor attempting to understand an organism without an understanding of these factors would be just as handicapped as the blind men were, feeling only one part of the elephant.

But an ecosystem is not just a plurality of interacting processes and functions through which organisms and communities affect one another in various ways. It is also a unity, a whole, as I am, in spite of my temporal and geographical interconnections; and this is true to some extent, whether or not one can tell precisely when and where it starts and ends. Of course, there is no claim made here that individuals and ecosystems are analogous in all respects. It simply appears that their status as developing and changing entities containing life renders them similar to each other in some relevant respects. It is their complex, changing, developing unity that is addressed by the concept of integrity.

From this discussion, several trends emerge that have persisted through the years and have become highly significant when the definitive definition of integrity was reached in 2000. We see the connection to climate change, the connection to soil erosion and, most important, the public health connection, inescapable from integrity's intimate relation to life itself. Also, the conceptual analysis of integrity we have described separates the integrity of single organisms from that of systems and cited the presence of possible different scientific methods to define integrity. We will briefly discuss these different methodologies together with the final definition.

6.2.2 *An Aside on the Situation We Face at This Time*

In the previous section, we noted that those who worked with the soil were fully aware of the connection between it and life as a self-generating force. They accepted that knowledge as present for decades. Why then do we need to conceptualize integrity as thoroughly as we can today, in order to present it to international legal fora? It is clear that in order to declare it the most viable starting point from which to attempt to redress the many policy errors that happened over time, based only partially on scientific mistakes but mostly on greed, we need to return to current scientific knowledge.

Politicians are constitutionally near-sighted; they only see as far as the next election. States and international organizations should share a much wider horizon instead. We have discussed why this does not seem to be the case. But the way forward, if at all possible, can only lie in the clear understanding of "all that science can still tell us" and acceptance of that direction. In 2019 climate change is a clear reality for all who can understand the messages now present in all media sources. Yet, even so, acceptance is still incomplete at best.

On December 7, 2019, the "State of the Union" short talk by Stefan Grober of Euronews, usually both informative and somewhat amusing, addressed climate change: the three countries most at risk were Germany, Japan and the Philippines, according to some scientific source he cited. For the Philippines, it is a clear case of extreme climate change events; for Germany, it seems, the earth was found to be as dry as the areas in North Africa, which no longer produce anything for the sustenance of the inhabitants, as noted in our discussion of migrants and refugees. Thus for Germany especially, but possibly at least in part for Japan as well (the latter is also prey to sea disasters), the presence of all the technical (chemical, etc.) tools of a rich country is impotent when faced with what I have defined as Hunger, so unless a radically changed system is devised for the provision of food, starvation may threaten even there.

On November 5, 2019, *BioScience* (American Institute of Biological Sciences) together with Oxford Academic published a "World Scientists' Warning of a Climate Emergency".[15] Right at the start, this document joins our main concerns: climate change and starvation:

> Profoundly troubling signs from human activities include sustained increases in both human and ruminant livestock populations, per capita meat production, fossil fuel consumption....[16]

The authors continue listing the "vital signs of climatic impacts":

> Three abundant atmospheric GHGs (C02, Methane, and Nitrous Oxide) continue to increase...as does global surface temperature....[17]

Based on their calculations, expressed in detailed tables in the article, the authors add:

> Despite 40 years of global climate negotiations, we have generally conducted business as usual...the climate crisis has arrived and is accelerating faster than most scientists expected...threatening natural ecosystems and the fate or humanity...potentially making large areas of the Earth uninhabitable.[18]

Further, in a short part titled "Nature", the authors voice their recommendations:

> We must protect and restore Earth's ecosystems...we need to quickly curtail habitat/biodiversity loss, protecting the remaining primary and intact forests...while increasing reforestation...at enormous scales.[19]

[15] Ripple, William J., Wolf, Christian, Newsome, Thomas M., Barnard, Pheebe, Mooman, William R., "World Scientists' Warning of a Climate Emergency", *BioScience*, Nov.5, 2019, bizz088: https://doi.org/10.109J/bioeci/biz088

[16] op. cit. p. 5.

[17] ibid. p. 6.

[18] ibid. p. 7.

[19] ibid. p. 9.

The question of food, highly significant in a document devoted to climate change, required "eating mostly plant-based food" while reducing "especially ruminant livestock, in order "to reduce methane emissions".[20] To my knowledge, no other movement, party or group has advocated essentially the same policies here recommended, other than what is now (and has been at least for the last twenty years) the Global Ecological Integrity Group. Our group, known as a "Project", was clear in its aim, that is, there was no wish to treat our work purely as a scholarly enterprise. But in order to be able not only to defend but especially to promote integrity, we had to start with a definition, a multidisciplinary unpacking of what it meant, so that we could then see its potential for change.

6.3 The Final Definition and the Essential Starting Point for the Decisive Employment of Integrity in Law and Public Policy

> Despite the bad press that generally follows an el Nino episode, on November 2, 1997, the Italian News Channel (RAI) and the US Sunday Report showed a marvel engendered by el Nino: the flowering of the Chilean desert….the unusual rains brought by el Nino produced a wonderland of flowers and grasses, with all accompanying complements or insects, such as bees, ants, butter flies, and other species.[21]

This passage is part of a chapter that leads to our concluding definition of integrity in 2000, by six co-authors. The basis or the phenomenon cited at the start of the section is no less than what the original understanding of integrity declared: when an ecosystem is largely unmanipulated, if not intact, it retains the biological potential and can follow its own evolutionary trajectory. But a suggestive passage is not sufficient to propose integrity as foundational for public policy, thus the need to join in a thorough understanding, reflected on a complete definition:

> The generic concept of integrity connotes a valuable whole, "the state of being hole, entire, undiminished, or "a sound, unimpaired or perfect condition".[22]

[20] Op.cit. p. 10.

[21] Laura Westra et al., 2000, "Ecological Integrity and the aims of the Global Ecological Integrity Project", in Pimentel, D., Westra, L., and Noss, Reed, *Ecological Integrity: Integrating Environment, Conservation and Health*, Island Press, Washington, DC, pp. 19–42, p. 19.

[22] *The Random House Dictionary of the English Language*, 1967.

At the start of the project, I had defined integrity as an umbrella concept that encompassed a variety of other concepts. In 2000, we started from a suggestive passage from wild nature, which still today provides the paradigmatic example of areas of integrity, untouched by industrial development. Another unique aspect of integrity was the autopoietic (self-creative) capacity of life to evolve, and, when anthropogenic impacts are largely absent, it is wild nature that shows itself to be the creation of "evolutionary and biogeographical processes at that place".[23]

At any rate, no reductionist representation of natural functions could possibly capture the fact that natural systems operate through their past and future, as nature's rhythms are displayed over time. No momentary snapshot captures all of nature's potential. The example of the Chilean desert is significative. In addition, conservation biology is concerned about the spatial requirements to maintain natural ecosystems, as systems of interacting and interdependent components need an open hierarchy of ecosystems.

Note, finally, that ecological integrity is valuable and valued. In the case of the Chilean desert, the dramatic transformation of "barren" desert into a vital and diverse biotic community provoked wonder and appreciation. Other ecological communities, such as reefs and rain forests, display their prolific life in a more continuous, less seasonal or episodic fashion. More generally, the biological and physical processes at work in these instances gave rise to the totality of life on Earth, including ourselves, and maintain the conditions for the continuation of life as we know it. Thus, natural ecosystems are valuable to themselves for their continuing support of life on Earth, as well as for the aesthetic value and the goods and services they provide to humankind. Indeed, ecological integrity is essential to the maintenance of ecological sustainability as a foundation for a sustainable society. For these reasons, there is a growing body of policy and law that mandates the protection and restoration of ecological integrity.

As a valuable and valued condition of biological systems, ecological integrity bridges the concerns of science and public policy. For both, we must be able to go beyond general qualitative descriptions to specify empirical and operational standards. Are integrity and its loss empirically measurable biological conditions? We believe so and present two basic approaches derived, on the one hand, from comparisons with a baseline condition in

[23] Engermeier, P.L., and Karr, J.R., 1994, "Biological Integrity versus biological diversity as policy directives", *Bioscience* 44, 690–697.

"wild" nature (i.e., places virtually free from human impacts) and, on the other, from complex systems theory. James Karr pioneered the creation of multimetric indices of biological condition (initially in streams and rivers) that measure the severity of biological degradation by deviations from a baseline condition of ecological integrity found in wild nature.[24]

Karr defines ecological integrity as "the sum of physical, chemical, and biological integrity".[25] Biological integrity, in turn, is "the capacity to support and maintain a balanced, integrated, adaptive biological system having the full range of elements (genes, species, and assemblages) and processes (mutation, demography, biotic interactions, nutrient and energy dynamics, and metapopulation processes) expected in the natural habitat of a region".[26] We can measure the extent to which a biota deviates from integrity by employing an IBI that is calibrated from a baseline condition found "at a site with a biota that is the product of evolutionary and biogeographic processes in the relative absence of the effects of modern human activity"[27]—in other words, wild nature. Degradation or loss of integrity is thus any human-induced position or negative divergence from this baseline for a variety of biological attributes.

Just as the index of leading economic indicators combines many financial measures to assess the state of the national economy, the IBI is holistic in that it integrates measurements of many biological attributes (metrics) to assess the condition of places. Metrics are chosen on the basis of whether they reflect specific and predictable responses of organisms to human activities. Ideal metrics should be relatively easy to measure and interpret. They should increase or decrease as identifiable human influence increases or decreases. They should be sensitive to a range of biological stresses, not narrowly indicative of commodity production or threatened or endangered status. Most important, biological attributes chosen as metrics must be able to discriminate human-caused changes from the background "noise" of natural variability. Human impact is the focus of biological monitoring and assessment.[28]

[24] Karr, James and E.W. Chu, 1999, *Restoring Life in Running Waters: Better Biological Monitoring*, Island Press, Washington, DC.

[25] Karr, James, 1996, "Ecological Integrity and ecological health are not the same", in P. Schulze ed., *Engineering Within Ecological Constraints*, Nat. Academy Press.

[26] Karr, James and Chu, E.W., 1995, "Ecological Integrity" Reclaiming Lost Connections", in L. Westra, J. Lemons eds. *Perspectives in Ecological Integrity*, Springer, The Netherlands.

[27] Karr (1996).

[28] Karr and Chu (1999).

Despite major variations in the structure and function of ecosystems throughout the world, a narrow range of IBI metrics has proven useful in evaluating the condition of places. IBI metrics evaluate species richness, indicator taxa (stress intolerant and tolerant), relative abundance of trophic guilds and other ecological groups, presence of alien species, or the incidence of hybridization, disease and anomalies such as lesions, tumors or fin erosion (fish) and head capsule abnormality (in stream insects).[29] Note that, unlike the procedures for standard water quality testing, physical attributes are biologically relevant; their impacts will be detected in the biological measures.

Regional calibration of IBI is appropriate because of natural differences between landscapes, just as different body temperatures are normal or healthy for different species of birds or mammals. As human actions touch almost all the different places on Earth, we have no choice but to attempt to understand the specific effect of these actions in each region. This requirement is not more stringent than the requirements of medicine, for instance. The appropriate diagnosis and treatment of disease may differ among infants, adults and senior citizens or among species treated with veterinary medicine.

Once a relevant standard and index of integrity have been established, various sites or areas can then be ranked by the extent of their deviation from the integrity standard or benchmark. IBI is a measure designed to document biological condition, the position of a site or landscape on a continuum from undisturbed (having biological integrity) to severely degraded. A heavily polluted or paved-over area where there is nothing alive has a biological condition marking an extreme of disintegrity, whereas the conditions of agricultural lands and commercial forest plantations lie near the middle of the spectrum. Only pristine or minimally influenced wild lands meet the integrity standard or benchmark. In effect, there are no significant degrees of integrity; it is a standard existing only at the top of the scale of the IBI. Although there is considerable pressure to invoke lower standards of evaluation, that integrity benchmark defined by wild nature should be retained so that citizens, politicians and policymakers know the level of degradation at sites subjected to the influence of human actions. Having a standard of biological integrity confers the ability to measure the full extent of our impacts on nature in exploited areas, as well

[29] Karr (1996).

as our success in protecting nature's legacy in wild areas. This is essential if we are to make informed social choices about land use.

Human survival depends on many of nature's "goods and services" that are invisible to markets and the economy; some are no doubt invisible to scientists. To know ourselves, we need to understand not only the processes and products of human history, culture and technology but also the processes and products of planetary evolution. We cannot hope to understand the effects of human actions on those products and processes without a systematic effort to evaluate trends in the condition of Earth's living systems.

Nevertheless, the Index of Biotic Integrity is not the only scientific approach on which our work was based. Sadly, our well-known colleague, James Kay, was deceased by 2000 (Kay was the scientist who actually hosted the first meeting, only a Focus Group of few people at the start in 1993, at the University of Waterloo, Ontario). Kay's friend and colleague, Robert Ulanowicz, worked on a complementary index of scientific performance, ascendancy, which assesses an ecosystem's performance at processing material and energy and thus helps to assess a system's ability to withstand stress. The overall indeterminacy of a system actually represents the system's developmental capacity or "the greatest possible(number) of developmental options within its time/location".[30]

The 2000 work also reprises the practical implications that follow upon our understanding, that is, the "practical steps" I defined "second-order principles" (SOPs) in my 1998 book, which proposed what was required by committing to "living in integrity".

How, then, can we conserve integrity and live sustainably? Westra (1998) proposes eight SOPs to define an ethics of integrity:

SOP 1 In order to protect and defend ecological integrity, we must start by designing policies that embrace complexity.

SOP 2 We should not engage in activities that are potentially harmful to natural systems and to life in general. Judgments about potential harms should be based on the approach of "post-normal" science.

SOP 3 Human activities ought to be limited by the requirements of the precautionary principle.

[30] Westra, op. cit., 1994.

SOP 4 We must accept an "ecological worldview" and thus reject our present "expansionist worldview" and reduce our ecological footprint.

SOP 5 It is imperative to eliminate many of our present practices and choices as well as the current emphasis on "technical maximality" and on environmentally hazardous or wasteful individual rights.

SOP 6 It is necessary for humanity to learn to live as in a "buffer". Zoning restraints are necessary to impose limits both on the quality of our activities and on their quantity. Two corollary principles follow: (a) we must respect and protect "core"/wild areas; (b) we must view all our activities as taking place within a "buffer" zone. This is the essential meaning of the ethics of integrity.

SOP 7 We must respect the individual integrity of single organisms (or micro-integrity), in order to be consistent in our respect for integrity and also to respect and protect individual functions and their contribution to the systemic whole.

SOP 8 Given the uncertainties embedded in SOPs 1, 2 and 3, the "risk thesis" must be accepted, for uncertainties referring to the near future. We must also accept the "Potency Thesis" for the protection of individuals and wholes in the long term.[31]

6.4 From Integrity's History to the Development of Policy for Global Governance

How do we strike a balance between fear and hope?[32]

That is indeed the question we need to ask ourselves. In 2019, the great majority of our group were legal scholars, some scientists, and the interest in public health is always present. So, the question where do we go from here involves, from the start, the strong critique of weak, ineffectual laws coupled with a strong desire to ensure that change can be introduced precisely at the level where it is most needed. This project will clearly not be one that can be promoted without encountering grave difficulties. Hence

[31] Westra, L., 1998, Living in Integrity, Rowman Littlefield, Landham, MD.

[32] Diamond, Jared, Apr.15, 2019 discussing Bill McKibben's book *Falter* (Has the Game Begun to Play Itself Out?), viewed Dec.9, 2019 at https://www.nytimea.com/2019/04/15/books/review/bill-mckibben-falter.html

our first step will necessarily be to assess the opposition we will encounter; an expected veritable war will have to be waged against corporate power.

Now as then, no world executive power exists. As a result, at the world level, a huge gap exists between the solemn statements of principle and bleak daily reality. The violation of human rights, conditions of extreme poverty, periodic recourse to war and environmental degradation are but a few of the many problems facing humankind today.[33]

Carl Boggs's point about "colonizing corporations",[34] while correct, does not go far enough today. Indeed, citizens rightly place the blame on corporate power, as they see dominant economic interests replacing any possible social interest that might move them to return to political concern and dialogue. But, after 2000, the most dangerous development yet in the growing trend of overwhelming, ubiquitous corporate presence is the infiltration of that power right into the government elites intended to protect the citizens, via the presence of "embedded science". These corporations now have complete confidence in their immunity from the legal consequences of any and all crimes they might commit, especially after the advent of the Terrorist Act of 2006, where the practical identification of large corporations with the US government itself is clearly enshrined.

The huge spread of corporate power transcends the global limits as well as those of the US domestic Constitutional law.

Daniele Archibugi notes the vast gulf between the affirmation of principles and the bleak—and I would add, paraphrasing Philippe Sands—"lawless" reality we face. He proposes "cosmopolitan democracy", based loosely on the West's vision of democracy. He proposes a cosmopolitan democracy transcending the democracy he believes exists in the West. He says, "The West in an entity composed of countries that have a market economy and consolidated democratic institutions. With the sole exception of Japan, the West involves Europe and its ancient settlements".[35] He thinks that—in addition—beyond the material power, we can observe the domination of the US and its ideology. Yet "freedom and democracy" are no longer the values that once carried basic universal connotations. Increasingly, these high-sounding concepts "have been turned into ideological screens to defend vested interests and attack enemies".[36]

[33] Archibugi, Daniele, 2008, *The Global Common Wealth of Citizens*, Princeton University Press, Princeton, NJ, p. 20.

[34] Boggs, Carl, 2000, *The End of Politics*, Guilford Press, New York, p. 119.

[35] Archibugi, D., op. cit. p. 3.

[36] ibid.

Hence that "world power" ought not to be the sole (or the greatest) source of power as often the West "claimed to have values (e.g. freedom and democracy) for itself and denied them to others".[37] Democratic regimes are not the only ones unworthy of carrying on the torch of civilization, as autocratic regimes may be equally, or even more likely to be, aggressive and to disregard human rights.

This is not the appropriate place to follow Archibugi's sophisticated analysis of all aspects of the possible achievement of a real global democracy, while evolving "meta-state law".[38] But we should consider the effect on "the West" of corporate persons, given the power these ubiquitous legal entities possess in the international community. Archibugi is correct as he says that "no world executive power exists", at least if he refers to an officially sanctioned or elected "executive power".

The questions we are raising in this work refer to the possibility that a supranational power does exist, especially in the West, but ultimately affecting the whole global community, which does not represent either "democracy" or "freedom". This power does not possess a true political ideology or a democratic mandate: it is only unashamedly limited to the aggressive pursuit off its own interests. As we continue to examine the corporate legal persons' power, together with their influence on public policy globally, we should keep in mind what institutions and organizations might have a legitimate claim on a true cosmopolitan democracy instead.

The US is governed by a type of "democratic elitism" but the secret institutions of the corporate directed second constitution "have marked antidemocratic authoritarian tendencies".[39] Miller does not discuss the actual detailed results of this duality, but he does state that the separation between the economy and the polity is only functional, as they are complementary instead. Hence a truthful understanding of the situation would be to term it the "corporate state".

This intimate relation entails that the most important aspects of democratic citizenship are eliminated. Mark Kesselman notes:

[37] ibid. p. 5.
[38] ibid., p. 228.
[39] Mill, Arthur, S., 1987, "Corporations and our two Constitutions", in Warren Samuels and Arthur Miller eds., Corporations and Society, Greenwood Press, New York, pp. 241–262.

> The hidden face of power is exercised not so much by the suppression of specific issues from the political agenda as by the exclusion of the most fundamental matters of public concern from the political sphere.[40]

The corporate control here acknowledged could not persist without an additional control, one we have neglected in this work given its wide reach and the public awareness of its results: the corporate control of the mass media. That is the final form of supranational control: neither the control of the law, of governance, nor of the markets exerts such widespread power. It is the ultimate social control of people's wants and preferences, so that unhealthy and even injurious choices become the "real" preferences of US citizens to begin with, but of people everywhere, eventually, as the "better way" is spread, from harmful McDonald's hamburgers and Coca-Cola to mind warping computer games and other such "necessities", promoted to the public from childhood on.

Thus, corporate control is not simply tolerated and accepted by the law. In contrast, again primarily in the US, the home base of most MNCs, the identification of corporate interests and legal regimes completes the structure of a "lawless world",[41] which explains and amply justifies public apathy and disinterest in what is left of "politics", while the mass flight to other, more reliable social movements and groupings attract the interest and allegiance of disenchanted transnational citizens, trying to recapture elsewhere the significant public "good" that has long since eluded them in their Western democratic national homes.

Politics, "in modern society", is conceived "in terms of elite action", and "the result was an impoverishment of the public sphere and political discourse that lingers on in many parts of the world even today".[42] The withdrawal of the states, both as a purveyor of public goods and as a reference of traditional principles and national values, left the citizens adrift. The void created was luckily filled by various social movements, presented and promoted through the internet. However, like the rest of the plethora of information that is found in the internet, not everything there is either truthful or valuable, and it is extremely hard to discriminate among the multiple offerings one can find, in order to be sure that one is truly in

[40] Kesselman, Mark, 1982, "The Conflictual Evolution of American Political Science: From Apologetic Pluralism to Dilate Realism and Marxism" in J. Greenstone ed., Public Values and Private Powers in American Politics, University of Chicago Press, Chicago, p. 571.

[41] Sands, Philippe, 2006, *Lawless World*, Allen Lane, London.

[42] Boggs, op. cit., p. 118 ibid.

touch with a principled organization, rather than yet another "green-washed" or otherwise altered version of the same corporate colonizers the citizen is attempting to escape.

Carl Boggs deplores the absence of political debate in the US, where "political colonization"[43] has taken the place of real debate between the parties that are almost indistinguishable, except perhaps from the respective amounts of the corporate funding they manage to attract.

While Archibugi takes a universal, principled position for global governance, Kesselman extends his reach to several voices in the political arena, including Marx. Nevertheless, both are strongly critical voices that originate from the West, thus from wealthy countries. We also need to turn to a strong voice from the East or South, one that is not related to the corporate West: Vandana Shiva.[44] Particularly telling is Chapter 3 of that work, titled "Bees and Butterflies Feed the World, Not Poisons and Pesticides", which clearly defends the natural cycles of ecologically biodiverse system as it appeals to Rachel Carson's seminal work.[45] Shiva remarks that the use of pesticides has increased exponentially over the last four decades. The corporations that produce pesticides are also in the business of producing pharmaceuticals and seeds; hence they market their products as "safe medicines for plants" and food for humans.[46] Yet these pesticides not only have their origin in chemical warfare, but their pesticides "do not control pests, rather they create them".[47]

Shiva reminds the reader that I.G. Farben (which included also BASF and Hoest), in the 1920s and 193Os, screened Zyklon B for Hitler's extermination camps.[48] From a legal standpoint, we can add that the Tribunals at Nuremberg after WWII condemned I.G. Farben/Bayer for that work. He was sentenced to spend seven years in jail, after which he emerged once again as a corporate hero of sorts, as Bayer placed Fritz ter Meer in an important position on their executive board.[49]

[43] Boggs, op. cit., ibid.

[44] Shiva, Vandana, 2016, Who Really Feeds the World? The Failures of Agribusiness and the Promise of Agriecology), North Atlantic Books, Berkeley, CA.

[45] Carson, Rachel, 1962, Silent Spring, Houghton Mifflin, Boston.

[46] Shiva, op. cit., p. 28.

[47] ibid.

[48] ibid. p. 29.

[49] Westra. Laura, 2018, *On Hunger*, International Press, Irvine, CA; see discussion of the issue on pp. 220–221.

Most of the 1400 pesticides in existence today are based on organophosphates (OPs), used as nerve gas in Nazi concentration camps, thus acting on the nervous system.[50] The devastating effects of poisons in the human diet are clearly in evidence in India, where nearly 700,000 Indians die of cancer every year. That figure rises to 8.3 million people globally according to the WHO's records for 2012.[51] Other countries in the global South are no better. Argentina produces soy using Roundup Ready, and in the "GM belt" of that country, both cancers and birth defects are prevalent. Finally, Brazil is the world's largest consumer of pesticides since 2008, and GM crops make up 45 percent of all crops in that country. Shiva concludes (and we concur):

This is not food production. This is war.[52]

These figures and that reality are not diminishing, but they are constantly on the increase today. As we suggested, we needed to start by knowing the enemy, in order to fight successfully. The next question is how best to frame that fight. On that, many suggestions exist, and they tend to line up in two basic camps. Some claim the only solution is to radically change the way society functions in the world. Others propose radically changing the laws to which most of us are already committed. We will discuss the second option in the final section of this chapter: let the reader decide which approach he or she finds more convincing.

6.5 The Radical Review of Legal Regimes for Integrity: Present and Future Paths

> Representing the commons, challenging capitalist accumulation, reshaping the law from the bottom up, and eventually changing the whole of our common understanding requires a robust theory and the daily widespread political resolve of individuals.[53]

I propose the work of the Global Ecological Integrity Group as the most appropriate means of "reshaping the law", something to which we have been committed from the start. Neither I nor those working with

[50] Shiva, op. cit, p. 31.

[51] Shiva, op. cit., p. 32, "Cancer Fact Sheet N. 297", WHO, February 2014, www.who.int. mediacentre/faotsheet/fs297/en/

[52] Shiva, op. cit., p, 37.

[53] Frittjof Capra and Ugo Mattei, 2015, The Ecology of Law, Barret, Koehler Publishers, Sn Francisco, CA, p. 156.

ecological integrity have any knowledge of how to radically reshape society as a whole. Also, that project, I believe, would be neither easy nor quick to accomplish, even if it were entirely feasible as many believe. That is a grave problem, when the world's situation in regard to both climate and food scarcity is so urgently in need of redress.

Revising legal regimes is not a quick fix by any means, nevertheless it is a system (legal regimes) to which both states and individuals are already committed, at least on paper, so it might be a project somewhat easier to promote. Others have proposed several possible approaches before my own and, for the most part, complementary to it. The first such theory we will review will be that of Mary Christina Wood:

> If indeed the body of public trust law aims to protect society and safeguard the natural infrastructure necessary for the citizens' survival and prosperity, courts can no longer tether it to artificial categories defined by courts two centuries ago. Nature's Trust recognizes a full ecological res. Ecological integrity rather than navigability, forms its talisman.[54]

The approach suggested by Mary Christina Wood is one with which I concur fully, although I might perhaps disagree with her exclusive reliance on state trust doctrine. Unlike the position of Weston and Bollier on "the commons", which included "informal, socially negotiated values, principles and values a given community develops", thus including much that is beyond the "natural objects" Wood's position that governance should be governed by the laws of nature, a critique that is well taken.[55]

In fact, Wood's argument in support of *Nature's Trust* is well worth a detailed discussion, because it offers a North American counterpart to the international ideal I propose as a basis for governance. The starting point is what Wood terms "the Illusion of Environmental Law", intended to protect the environment, but instead it permits environmental agencies "to legalize destruction of the environment", as "most agencies today" are aligned "all too closely with the industries they regulate". In addition, "the judiciary has largely relinquished its role as an institutional check on environmental agencies, regularly invoking the administrative defense

[54] Wood, Mary Christina, 2014, *Nature's Trust*, Cambridge University Press, New York.

[55] Weston, Burns H. and Bollier, David, 2013, Green Governance: Ecological Survival, Human Rights and The Law of the Commons, Cambridge University Press, NY, p. 104.

doctrine". However, "environmental law scholarship ignores these systemic problems".[56]

As we saw, the best legal scholars address each problem in isolation, while new and emerging legal instruments are—at best—procedural in content, while the old "notify-permit-destroy cycle of environmental law" persists: that must be the first focus of radical change. That is the main focus of integrity as well, as all systemic components, biotic and abiotic, function together, and the removal of even one species may have grave and unexpected consequences.[57] Ecosystems include physical and chemical components, such as soils, water and nutrients that support the organisms living within them. These organisms may range from large animals and plants to microscopic bacteria. Ecosystems include the interactions among all organisms in a given habitat.

The clearest example of trust principles enacted by governments can be found today in their legal position regarding Indigenous peoples, Native Americans in the US, First Nations in Canada, Aboriginals in Australia. In Canada, the fiduciary duty of the Crown is undisputed, and it and it alone trumps even economic interests in the case law regarding their lands and traditional pursuits.

The public trust, beyond police powers and regular statutes, can indeed "limit the scope of government action":

> A sovereign property interest arises from government's control over a particular territory (just as private title to property comes from an individual's association with a particular parcel). The trust defines the character of this sovereign property ownership, recognizing rights held perpetually by the people. Weaving together two civic strands of power and obligation the trust embodies: (1) the people's delegation of authority to their government to control and manage resources; and (2) the people's assertion through a fiduciary obligation of limits on that authority to ensure that it functions to benefit the public rather than special interests (who may have greater sway over legislative process.[58]

Thus, not "public trust" in general, but the ecological trust specifically, as it unites Wood's understanding of the concept and the ongoing fiduciary obligation of states should be adopted. In fact Wood specifically addresses

[56] Wood op. cit. pp. 9–10.
[57] ibid. pp. 119–120.
[58] Wood, op. cit., p 129.

the "ecological res", that is, the ecological assets, the natural wealth that must sustain all foreseeable future generations of humanity.[59] In contrast, the protection of integrity, while recognizing the role of ecosystems, nature itself, in human survival, also proposes respect and concern for all forms of life, for their intrinsic value as well as for their role in ensuring human survival.

However, Wood is certainly correct in emphasizing human survival as a non-derogable obligation of states, and as the basis for "public concern", so that public trust should not "surrender to statutory law in case of conflicts".[60] The major differences between public trust and statutory law lies in the fact that although common law is intended to be both flexible and adaptable, special interests render it either flexible enough to fit such interests or, at times, too fixed to incorporate the changes necessary to reflect ongoing scientific discoveries.

And here lies the crux of the problem, whether we consider "nature's trust" or whether we want to protect biological integrity: all human enterprise requires use, and often privatization, of some areas of nature, so that all forms of commodification cannot be proscribed. But, starting from the current "scientific understanding" is basic to both nature's trust arguments and those in support of principle of integrity. Wood cites the Ecological Society of America:

> Ecosystems include physical and chemical components, such as soils, water and nutrients that support the organisms living within them. These organisms may range from large animals and plants to microscopic bacteria. Ecosystems include the interactions among all organisms in a given habitat.

Thus, once science established equally the relation between human survival and ecosystem services, it is only the perverse clinging to the industrial/economic interests arising from globalization that can allow the ongoing overconsumption of "resources", the plunder and destruction that govern today's permissible, legal corporate activities. In addition, the approach employed by regulatory regimes is—at best—an attempt to "save one species or another at a time", "evaluating each asset in isolation to determine whether it merits protection".[61] However, the opposite is

[59] Wood, op. cit. p. 143.
[60] ibid. p. 149.
[61] Capra and Mattei, op. cit. p. 179.

necessary, as I argue in defense of integrity, that is, of the wholeness of each functioning system.

At any rate, the differences between Wood's understanding and ours in this work are not too grave: her reference to the unique and irreplaceable common heritage though more limited is another connection. Before turning to a further consideration of the differences between Wood's worth and mine, we need to note her basic plan for the trustees:

> The six substantive fiduciary duties require a public trustee to: (1) protect the res; (2) conserve the natural inheritance of future generations (the duty against waste); (3) maximize the societal value of natural resources; (4) restore the trust res where it has been damaged'; (5) recover natural resource damage from third parties that have injured public trust assets; and (6) refrain from alienating (that is privatizing) the trust, except in limited circumstances.[62]

Wood's approach is close to our own, although the major difference is that we acknowledge the inherent weakness of states, and we have had reseed this issue already. Most of the problem rests with the corporate power which now operates as a supranational entity, bypassing the various states whose power it has already usurped. Therefore, relying on the states' support of ecological integrity, even if a declared objective, may lie beyond what these states are capable of achieving. Thus, I regretfully believe that, while it is useful to target single states in the hope of awakening them to their responsibilities, that plan may not be enough.

Klaus Bosselmann, who works with the integrity group, proposes another form of ecologically based governance, one based on trusteeship:

> In summary, ecosystem services are vital for human survival and well-being. Biodiversity in turn is vital for healthy ecosystem functioning. Yet biodiversity if under unprecedented threat as the result of thoughtless, unknowing human activity. We are determining the very support structures for out biological survival and cultural vitality.[63]

Economists, ecological economists, activists, conservation biologists, legal scholars and even Pope Francis appear to agree on this assessment of

[62] Wood, op. cit. p. 115.

[63] Sachs, Jeffrey O., 2016, The Age of Sustainability, Columbia University Press, NY, p. 452.

the current situation, regarding both this issue and climate change. Thinking people everywhere agree on the diagnosis, while still holding highly divergent positions on the possible cure or the solutions that might work.

Each time we discussed one of these positions, we acknowledged the difficulties that accompanied each one, and we'll return to a summary of these difficulties in the next section. For now, we need to focus on the positive aspects present in each strategy, although none of them appears to offer a solid solution as its stands. Thus the question that arises is, leaving aside for now the problems and obstacles that plague each proposal, what is the most promising aspect of each: the first is the rediscovery and revaluing of the commons, the "tragedy" of which started, with the enclosure movements in England, but gained an important rebirth recently. Klaus Bosselmann says:

> The point is to recapture the narrative about our shared wealth in the community of life and shed the instrumental approach to nature, by including more ecological, social, democratic and moral value.[64]

He adds, speaking of "tragedy of the commons": "Rational self-interest gives rise to an irrational outcome". When we consider the rise to attention of the concept of the commons, as we see in the Charter of the Forest, it included not only the necessary space for grazing sheep but the forest, the woods, the rivers, the peasants' abodes and the animals in "the commons" or rather, as it was called at the time, the "forest".

One wonders why, if an absolute medieval monarch could be persuaded that he could not exploit "the forest" for himself and have guards and "beadles" to punish trespassers or those who questioned his supreme power over those aspects of his domain, today's democratic states cannot seem to understand the urgency of that people's needs and their rights. The reason for this lack of understanding is that any nobleman or king at least realized that peasants had to be kept alive and fit to work, minimally, even if he had little concern for their specific interest.

In contrast, the only concern of today's Western states and the corporations that support them is to increase their joint economic and geographical power. The physical fitness of their workers is no longer all-important, because the functions they perform, for the most part, require cleverness, technical expertise, not brawn. Further, should they become ill because of

[64] Bosselmann, Klaus, 2015, Earth Governance, Edward Elgar Publishing, Cheltenham, p. 53.

the lack of a healthy/safe environment, some major powers have no social medicine, and in most countries, citizens must purchase medicines at the outrageous prices Big Pharma enforces, often for diseases they could have avoided, if appropriate concern for the state of the commons had been practiced.

The second positive strategy we noted was that of viewing states as "trustees", not sovereigns controlling the natural commons in their own interest, rather, holding them in trust for the public good, the common interest of humanity, at least each state for their own citizens. We will return to this concern below, as it is just an aspect of an otherwise excellent ecological strategy.

Bosselmann expresses this concern as well: "the appetite of states to exploit the global environment is greater than the interest in preserving it".[65] Yet the strategy of trusteeship, proposing a fiduciary relation to what exists presently between state and citizens, has a powerful appeal. It seems obvious that the allegiance of citizens, at least insofar as carrying their country's passports, embracing the main obligations of citizenship (paying taxes, obeying the law, respecting police and state functionaries), has as its counterpart the state's own obligation to protect. However, it is precisely the meaning and extent of that protection that is and remains in question.

Unquestionably, the most significant aspect of the trust doctrine is to recall explicitly governments to their obligation to citizens, as the protection for their life is the governments' first obligation, an obligation not often acknowledged in the jurisprudence. The obligation to protect seems to be indisputable and easy to support in law, although its reach and full meaning is in question and may bleed some serious re-examination.

We are committed to reaching a regime that would support global ecological integrity; hence any legal regime that is limited to a state's own citizens is at best necessary, but not sufficient for our purpose.

Hence what is necessary is to apply the positive aspects of "nature's trust", as it applied to states, to a global situation. Wood asserts that "once a natural resource becomes identified as a planetary trust asset, legal principles evaluate its management according to measures of asset health and functionality", which she views as basic, a first step so that "the trust formulation creates a uniform and principled approach to global ecological responsibility". Such a situation should ensure that "citizens should

[65] Bosselmann, op. cit. p. 186.

pursue enforcement of global trust obligations against their own governments in domestic courts".[66]

We will consider below whether such a suggestion is feasible, based as it is on a "holistic approach to defining government obligation". In fact, it is this "holistic approach" that led to a further consideration of the common heritage of mankind (CH) principle not presently applied to global ecological responsibility, but perhaps representing the best potential for such a novel application of an existing regime.

The other important aspect from the standpoint of ecological integrity is the fact that a proposal to the UN, possible development of the binding instrument, is the result of the considerations of a learned group of scholars, not simply of an open "democratic" vote.

In fact the consideration of the common heritage of mankind also presents an excellent approach to our aims, particularly because of their 2015 presentation to the United Nations.[67] Aside from the great significance of this UN presentation, the appeal to oceans' role regarding climate change is basic, as we shall see below from a 2019 report.

A recent article reports the role that the presently degraded oceans could play regarding climate change[68]:

> Halting overfishing and the plastic pollution of the oceans could help tackle the climate emergency by improving the degraded state of the world's biggest carbon sink, a report has found.[69]

The present problems include our overconsumption of marine life and the overwhelming presence of plastics in the oceans, which limit drastically the oceans' absorptive capacities. The report proposes creating "ocean sanctuaries" and forging a new treaty for their protection. This is what, in our original work, was termed creating buffer zones to protect main

[66] Wood, op. cit. p. 219.

[67] on January 23, 2015, an Ad Hoc Working Group to the United Nations, regarding a possible future third UNCLOS Implementation Agreement; see Tullio Scovazzi, 2015, "The Conservation and Sustainable Use of Marine Biodiversity, Including Genetic Resources in Areas Beyond National Jurisdiction: A Legal Perspective", available at www.un.org/Depts/los/consultatlveprocess/ICP12_Preaentations/Scovazzi Presentation.pdf

[68] Harvey. Fiona, in Madrid, "Seascape: The State of our Oceans" supported by the David and Lucile Packard Foundation; "Tackling degraded oceans could mitigate climate crisis", Report (viewed Dec. 9, 2019 at https://www.theguardian.com/environment/2019/de/04/tacklingoceans

[69] op. cit. p. 2.

protected areas of integrity in ecosystems. In the oceans, algae transform dissolved carbon dioxide into organic carbon, which then becomes part of the food chain. In addition, krill (small crustaceans) play a role as well, feeding larger species, but they too are declining because of overfishing and climate change. Protecting the oceans not only would help reduce climate change in general, but it would particularly help the impact of rising sea levels on islands and ocean-side nations.[70]

Hence several themes proposed by the ecological integrity group right from the start are present in this report: first, the ecosystem approach is taken for granted; second, the need for protected areas of integrity, together with buffer zones to further isolate them from possible harmful activities; and third, the return in some sense to our starting point, that is, the integrity of water, for which we can remember the successful approach to eliminating harmful algae and other non-native species from Lake Erie in Canada which, although technically dead in the 1980s, eventually returned to life (i.e., to be normally stocked with fish).

Another point worthy of note is that overfishing entails the overuse or a diet that recommends some fish but also depends on vegetables and legumes for the most part of our diet, that is, the recommended Mediterranean diet.

We have now come to the conclusion of this work, by stating outright the other possible position to promote in order to defend against both climate change and starvation: the radical change and development of international law. The research of many years has indicated the damage legal positivism, on one side, and corporate power, on the other, were doing to legal regimes, ostensibly created to protect justice, human rights and the public order. In fact, even the most entrenched international legal regimes are now treated by clever lawyers as something to circumvent and use to ensure that the expected protection of the weak, the poor, would not take place.[71]

International law does have all the principled instruments needed to stop state inaction/complicity and corporate intent to advance its own interests against the interests and safety of humanity.[72] Especially recently

[70] op, cit. p. 3.

[71] Erakat, Noura. 2019, *Justice for Some (Law and the Question of Palestine)*, University of Stanford Press, Stanford, CA.

[72] Westra, Laura, 2019, *From Principles to Norms: The Development of International Law*, Nova Science Publishers, New York.

many scholars have noted this grave problem over the last decade, and I have treated the issue in detail availing myself of their work. This is not the appropriate time to return to that material. However, there is a short passage from a major scholar in law that has been my inspiration for my last book that I would like to cite. It is by a well-known law professor, Antonio Cassese: his work and that passage were the foundation not only for that book but also for the way I hope my work and that of my group will proceed, a starting point for our future endeavors. Cassese is listing the many international legal instruments designed to defend basic principles, and then he adds:

> The situation presents a remarkable oddity, which however is indicative of the still rudimentary development of international law, on the one hand there exist fundamental principles which comprise the "international public order", principles from which consequently States cannot derogate in their dealings, on the other hand it is only possible to rely upon these principles in relatively exceptional circumstances. Such principles thus remain in a state of potentiality, rather than producing their legal effects on an everyday basis and in any direction.[73]

Indeed, the law's development is still only "rudimentary" today, as it was in 2005. The fundamental principles to which Cassese refers are there, present, but in order to protect economic and power related interests, their use is circumvented, and the potentiality of international law cannot emerge and flourish. I hoped that my 2019 book could be accepted as the first step toward that development and that flourishing. I trust that this work will also represent another useful step in that direction.

References

Archibugi, Daniele, 2008, *The Global Common Wealth of Citizens*, Princeton University Press, Princeton, NJ, p.20

Boggs, Carl, 2000, *The End of Politics*, Guilford Press, New York, p.119

Bosselmann, Klaus, 2015, Earth Governance, Edward Elgar Publishing, Cheltenham, p.53

Capra, Frittjof and Ugo Mattei, 2015, The Ecology of Law, Barret, Koehler Publishers, Sn Francisco, CA, p.156

[73] Cassese, Antonio, 2005, International Law 2nd ed., Oxford University Press, Oxford, p. 204.

Carson, Rachel, 1962, Silent Spring, Houghton Mifflin, Boston
Cassese, Antonio, 2005, International Law 2nd ed., Oxford University Press, Oxford, p.204
Christian Farmers Federation of Alberta and Ontario, *Earthkeeping*, Vol.6, No.3, March, 1991
Diamond, Jared, Apr.15, 2019 discussing Bill McKibben's book *Falter* (Has the Game Begun to Play Itself Out?), viewed Dec.9, 2019 at www.nytimea.com/2019/04/15/books/review/bill-mckibben-falter.html
Efron, Edith, 1984, *The Apocalypticsl PoliticsıScience and the Big Cancer Lie*, Simon and Schuster, New York
Engermeier, P.L., and Karr, J.R., 1994, "Biological Integrity versus biological diversity as policy directives", *Bioscience* 44, 690–697
Erakat, Noura. 2019, *Justice for Some (Law and the Question of Palestine)*, University of Stanford Press, Stanford, CA
Genesis 1:31a
Harvey, Fiona, in Madrid, "Seascape: The State of our Oceans" supported by the David and Lucile Packard Foundation; "Tackling degraded oceans could mitigate climate crisis", Report (viewed Dec. 9, 2019 at https://www.theguardian.com/environment/2019/de/04/tacklingoceans
Karr, James and Chu, E.W., 1995, "Ecological Integrity: Reclaiming Lost Connections", in L. Westra, J. Lemons eds. *Perspectives in Ecological Integrity*, Springer, The Netherlands
Karr, James and E.W. Chu, 1999, *Restoring Life in Running Waters: Better Biological Monitoring*, Island Press, Washington, DC
Karr, James, 1996, "Ecological Integrity and ecological health are not the same", in P. Schulze ed., *Engineering Within Ecological Constraints*, Nat. Academy Press
Kesselman, Mark, 1982, "The Conflictual Evolution of American Political Science: From Apologetic Pluralism to Dilate Realism and Marxism" in J. Greenstone ed., Public Values and Private Powers in American Politics, University of Chicago Press, Chicago, p.571
Mill, Arthur S., 1987, "Corporations and our two Constitutions", in Warren Samuels and Arthur Miller eds., Corporations and Society, Greenwood Press, New York, pp.241–262
Norton, B.G., 1986, Why Preserve Natural Variety? Princeton University Press, Princeton, NJ, pp.49–60
O'Neill, Robert, DeAngelis, D. et al., 1986, The Hieratical Concept of Ecosystems, Princeton University Press, Princeton
Odum, E. P 1993, "The Ecosystem", *Ecology*, 2nd ed., Sinaver Association, Sunderland MA
Pope Francis, 24 May 2015, "Encyclical Letter, Laudato Si' of the Holy Father Francis on Care for Our Common Home", available at http://w2.vatican.va/content/francesco/en/encyclicals/documents/papa-francesco_20150524_encyclica-laudato-si.html

Ripple, William J., Wolf, Christian, Newsome, Thomas M., Barnard, Pheebe, Mooman, William R., "World Scientists' Warning of a Climate Emergency", *BioScience*, Nov.5, 2019, bizz088: https://doi.org/10.109J/bioeci/biz088
Rolson, Homes III, 1988, *Environmental Ethics*, Temple University Press, Philadelphia
Rosen, Robert, 1991, *Life Itself, A Comprehensive Inquiry into the Origin, Nature of Life*, Columbia University Press, p.116
Sachs, Jeffrey O., 2016, The Age of Sustainability, Columbia University Press, NY, p.452
Sands, Philippe, 2006, *Lawless World*, Allen Lane, London
Scovazzi, Tullio, 2015, "The Conservation and Sustainable Use of Marine Biodiversity, Including Genetic Resources in Areas Beyond National Jurisdiction: A Legal Perspective", available at www.un.org/Depts/los/consultatlveprocess/ICP12_Preaentations/ScovazziPresentation.pdf
Shiva, Vandana, 2016, Who Really Feeds the World? (The Failures of Agribusiness and the Promise of Agriecology), North Atlantic Books, Berkeley, CA
The Random House Dictionary of the English Language, 1967
Weston, Burns H. and Bollier, David, 2013, Green Governance: Ecological Survival, Human Rights and The Law of the Commons, Cambridge University Press, NY, p.104
Westra, Laura et al., 2000, "Ecological Integrity and the aims of the Global Ecological Integrity Project", in Pimentel, D., Westra, L., and Noss, Reed, *Ecological Integrity: Integrating Environment, Conservation and Health*, Island Press, Washington, DC, pp.19–42, p.19
Westra, Laura, 1994, *The Principle of Integrity*, Rowman Littlefield, Lanham, MD
Westra, Laura, 1998, Living in Integrity, Rowman Littlefield, Lanham, MD
Westra, Laura, 2018, *On Hunger*, International Press, Irvine, CA, pp. 220–221
Westra, Laura, 2019, *From Principles to Norms: The Development of International Law*, Nova Science Publishers, New York
Wood, Mary Christina, 2014, *Nature's Trust*, Cambridge University Press, New York

Index[1]

[1] Note: Page numbers followed by ‘n’ refer to notes.

L. Westra, *Climate Change and Starvation*, Palgrave Insights into Apocalypse Economics,
https://doi.org/10.1007/978-3-030-42124-3

The manufacturer's authorised representative in the EU is Springer Nature Customer Service Centre GmbH, Europaplatz 3, 69115 Heidelberg, Germany. If you have any concerns regarding our products, please contact ProductSafety@springernature.com

Printed and bound by CPI Group (UK) Ltd, Croydon, CR0 4YY
25/09/2026
02222898-0001